BRIGHT AIR,
BRILLIANT FIRE

GERALD M. EDELMAN

BRIGHT AIR, BRILLIANT FIRE

―

On the Matter of the Mind

BasicBooks
A Division of HarperCollins*Publishers*

Library of Congress Cataloging-in-Publication Data
Edelman, Gerald M.
   Bright air, brilliant fire: on the matter of the mind/ Gerald M. Edelman.
      p.   cm.
   Includes bibliographical references and index.
   ISBN 0–465–05245–2
   1. Mind and body.   2. Neuropsychology.   3. Philosophy of mind.
I. Title.
BF161.E34   1992   91–55454
128′.2—dc20   CIP

To the memory of two intellectual pioneers,
Charles Darwin and Sigmund Freud.
In much wisdom, much sadness.

For by earth we see earth, by water water;
by air bright air, and by fire brilliant fire

—Empedocles

And going on, we come to things like evil, and beauty, and hope . . .

Which end is nearer to God; if I may use a religious metaphor. Beauty and hope, or the fundamental laws? I think that the right way, of course, is to say that what we have to look at is the whole structural interconnection of the thing; and that all the sciences, and not just the sciences but all the efforts of intellectual kinds, are an endeavor to see the connections of the hierarchies, to connect beauty to history, to connect history to man's psychology, man's psychology to the working of the brain, the brain to the neural impulse, the neural impulse to the chemistry, and so forth, up and down, both ways. And today we cannot, and it is no use making believe that we can, draw carefully a line all the way from one end of this thing to the other, because we have only just begun to see that there is this relative hierarchy.

And I do not think either end is nearer to God.

—Richard Feynman

Contents

CONTENTS

PART IV
HARMONIES

List of Illustrations

of mind beginning in the sixth century B.C. It comes from a fragment written by Empedocles, physician, poet, and an early materialist philosopher of mind. His idea that perception results from the fit and size of material entities to particular pores in our bodies is more appropriate to modern theories of smell than of vision, but his heart (where he thought his mind was) was in the right place.

PART I

PROBLEMS

If we consider that without a mind no questions can be asked, and that there has never been a solidly established demonstration of a mind without a body, the importance of the subject addressed here needs no defense. In this part of the book I want to introduce the reader to some classical thoughts about the mind. I also want to hint at what is attempted later: to describe a biological theory of how we come to have minds. To do so I shall go into the organization of the actual matter underlying our minds—neurons, their connections, and their patterns.

CHAPTER 1

Mind

Cogito, ergo sum.

—René Descartes

The defect of Descartes' Discourse on Method lies in his resolution to empty himself of himself, of Descartes, of the real man, the man of flesh and bone, the man who does not want to die, in order that he might be a mere thinker—that is, an abstraction. But the real man returned and thrust himself into his philosophy. . . .

The truth is sum, ergo cogito—*I am, therefore I think, although not everything that is thinks. Is not conscious thinking above all consciousness of being? Is pure thought possible, without consciousness of self, without personality?*

—Miguel de Unamuno

"Don't think of an elephant."
Of course you did, and so did I. But where is the elephant? In your mind, and certainly not in the room, at least for most people who are reading this book. *Not* to think of it, you had to know what it was, to have remembered it, and even, in some cases, to have entertained an image of it. Above all, you had to know this language and how to understand this bit of wordplay.

Another piece of wordplay—"What is mind? No matter. What is matter? Never mind"—tells you the conclusion René Descartes (figure 1–1)

3

FIGURE 1–1

René Descartes (1596–1650), one of the founders of modern philosophy and a great mathematician. The dualism he espoused still polarizes modern thought about the mind. Cartesian dualism is likely to be dispelled only when we understand the relationship between consciousness and physics.

came to in his thoughts about the subject. Those thoughts marked the beginning of modern philosophy and they split the mind away from scientific inquiry. To Descartes, the mind was a special substance, one not located in space, not an extended thing the way matter was. This doctrine of dualism has plagued us—if not most of us, then at least many philosophers and some theologians—ever since.

What does it mean to have a mind, to be aware, to be conscious?

Everyone has thought about this at one time or another, but until recently scientists as scientists have shied away. Now there is something new on the scene: neuroscience. We have begun to accumulate scientific knowledge about the brain at an explosive rate. It is becoming possible to talk in scientific terms about how we see, hear, and feel. The most complicated object in the universe is beginning to yield up its secrets.

Why should we think that this will tell us anything about our minds? Because of what we have already learned: just as we have recognized how matter comes to be in terms of particular arrangements of things, we should be able to figure out how minds arise out of other such arrangements. That is what this book is about: connecting up what we know about our minds to what we are beginning to know about our brains.

I will range over a variety of topics: nerves, computers, perception, language, selves. I will try to show how they connect both to one another and to our being aware. Rather than talk about how we think or reason, I am going to discuss the *basis* for these high-level activities. My overall goal is to show that it is scientifically possible to understand the mind. I will try to keep the technical details to a minimum, at the same time not hesitating to take on shibboleths and received ideas when I believe they are in error. Thus, although parts of this book will be concerned with pointing out positions I believe are indefensible, I promise that the main thrust will be positive and constructive. After all, the subject, like obstetrics, is well nigh indispensable to our being here. It is at the center of human concern.

Let us begin.

The word *mind* prompts thoughts of abstruse philosophical discussions. But it also casts a familiar shadow imported from everyday use—"What's on your mind," "Never mind," "Minding the baby." It is not at all clear what is being referred to in these expressions. But we can still rely on some commonsense notions to get started:

1. Things do not have minds.
2. Normal humans have minds; some animals act as if they do.
3. Beings with minds can refer to other beings or things; things without minds do not refer to beings or other things.

This last property, called intentionality by the German philosopher Franz Brentano, serves as a good indicator of the existence of a mental process. It refers to the notion that awareness is always *of* something, that it always has an object. I will refer to intentionality often in what follows.

But a set of indicators is not enough—we want to find out how the mind relates to matter, particularly to the special organization of matter that underlies it. It is not surprising that people have treated the mind itself as a special thing or a special form of stuff. After all, it seems so different from ordinary matter that its possessor may find it difficult to conclude by introspection alone that it could arise from the *interactions* of nonintentional matter. But as William James (figure 1–2) pointed out, mind is a process, not a stuff. Modern scientific study indicates that extraordinary processes can arise from matter; indeed, matter itself may be regarded as arising from processes of energy exchange. In modern science, matter has been reconceived in terms of processes; mind has not been reconceived as a special

FIGURE 1–2

William James (1842–1910), one of the founders of modern physiological psychology and an exponent of the philosophy of pragmatism. His thoughts on consciousness—that it is a process not a substance; that it is personal and reflects intentionality—shape much of our modern view of the subject.

form of matter. That mind is a special kind of process depending on special arrangements of matter is the fundamental position I will take in this book.

If we look at the commonsense list we started with, we see that biological organisms (specifically animals) are the beings that seem to have minds. So it is natural to make the assumption that a particular kind of *biological* organization gives rise to mental processes. Obviously, then, to pursue the subject scientifically we must turn to how the brain is organized. It would be a mistake, though, to ignore the rest of the body, because there is an intimate relation between animal functions (especially movement) and the development of the brain.

Since Darwin, biologists, when faced with particular kinds of biological organization, have almost automatically considered how evolution might have given rise to them. Brain and mind are no exceptions. Therefore we will also want to know something about how the brain structures underlying the mind arose in evolutionary history.

Above all, what we want to know is how such structures work. This is where advances in neuroscience come to the fore. It is exciting to contemplate the possibility of relating these advances to the accomplishments of psychologists studying behavior and mental processes. The findings of neuroscientists indicate that mental processes arise from the workings of enormously intricate brain systems at many different levels of organization. How many? Well we don't really know, but I would include molecular levels, cellular levels, organismic levels (the whole creature), and transorganismic levels (that is, communication of one sort or another). Each level can be split even further, but for now I will consider only these basic divisions.

It is startling to realize how many connections project from any one level to another—from a fear response induced by a warning cry to a biochemical process that affects future behavior; from a viral infection to a change in brain development that alters maturation; from a perception of a pattern to the chemistry of changes in a muscle; from any of these at some critical time of development to how a human child develops a self-image— strong or inadequate, detached or dependent.

To explain these kinds of changes, I first have to clear up some misconceptions. These have arisen mainly because experts in various subdisciplines have remained confined within their own specialties. But this is not the only reason. Prejudice, the inability to carry out certain experiments, and the traps of language have all made it difficult to tease out the connections between mental events and events in the nervous system.

There is more to studying the problem of mind than these matters of clarification indicate. As we will see, methods of doing science on inanimate objects, while fundamental, are not adequate to doing science on animals that have brains and possess intentionality. This is because scientific observers themselves are intentional animals, locked into their own experiences of consciousness, who must ensure that their observations can be communicated to other observers effectively, meaningfully, and without prejudice. This means they cannot include—indeed, they must deliberately exclude—elements in their own private experience or awareness. We can say this in a flurry of rhymes and near rhymes: intersubjective communication in science must be objective, not projective. No wonder that magic, vitalism, and animism pervaded prescientific communication. The projection of individual wishes, beliefs, and desires was not only allowed but was a major goal to be achieved in organizing societies for defense against natural threats in a sensible way.

None of this means, however, that a scientific study of the mind is impossible. It does mean that such a study will be full of pitfalls, hidden postures, and received ideas, many drawn from science itself. Even the most intelligent researchers working on the properties of mind have stumbled. And in studying intentionality, some persist in what can only be considered a parody of successful sciences like physics, sciences that are dedicated to the study of objects that lack intentionality.

How can we avoid falling into these traps? One way is to take the existing traps apart and ask whether modern neuroscientific research can help us dismantle them. Let us turn to these tasks, particularly to the task of putting the mind back into nature without abandoning scientific procedure.

CHAPTER 2

Putting the Mind Back into Nature

The way in which the persecution of Galileo has been remembered is a tribute to the quiet commencement of the most intimate change in outlook which the human race had yet encountered.
—Alfred North Whitehead

I want to go back to the beginnings of modern science and consider two towering figures of the seventeenth century, Galileo Galilei (figure 2–1) and René Descartes. In *Science and the Modern World*, Alfred North Whitehead observed that in inventing mathematical physics, Galileo removed the mind from nature. By this figure of speech, I suppose he meant that Galileo insisted that the observer must be objective, that he must avoid the vexing disputes of Aristotelian philosophers over matters of causation. A scientist should instead make measurements according to a model with no human projection or intention built into it and then search for correlative uniformities or laws that either support or disconfirm his or her claims.

This procedure has worked magnificently for physics and its companion sciences. Isaac Newton stands as the triumphant figure of its first full flowering. Even today after the Einsteinian revolution and the emergence of quantum mechanics, the Galilean procedure has not been swept aside. Albert Einstein's theory of relativity showed how the position and the velocity of the observer altered the measurement of space and time, and by taking acceleration into account it altered the very meaning of the word

9

FIGURE 2–1

Galileo Galilei (1564–1642), physicist, mathematician, and astronomer; the founder of a truly modern physics and, some would say, of modern science. In 1633 he was forced by the Roman Catholic Church to recant the heliocentric views espoused in his Dialogue on the Two Chief World Systems, *and spent the rest of his life under house arrest.*

matter. Quantum mechanics showed that the operation of measurement in the domain of the very small ineluctably involves the actions of the observer who has to choose, within the uncertainty dictated by Planck's constant, the level of precision with which he or she wishes to know either the position or the momentum of a subatomic particle. This reflects what physicists call the Heisenberg uncertainty principle.

Even with the startling revelations that at velocities approaching that of light or at very small distances the observer is embedded in his or her measurements, the goal of physics remains Galilean: to describe laws that are invariant. We have no reason to abandon this goal. This is because Einsteinian and Heisenbergian observers, while embedded in their own measurements, are still psychologically transparent. Their consciousness and motives, despite occasional arguments about their importance to quantum measurements by philosophers of physics, do not *have* to be taken into account to practice physics. The mind remains well removed from nature.

But as Whitehead duly noted, the mind was put back into nature with the rise of physiology and physiological psychology in the latter part of the nineteenth century. We have had an embarrassing time knowing what to do with it ever since. Just as there is something special about relativity and quantum mechanics, there is something special about the problems raised by these physiological developments. Are observers themselves "things," like the rest of the objects in their world? How do we account for the curious ability of observers (indeed, their compelled need) to carve up their world into categories of things—to refer to things of the world when things themselves can never so refer? When we ourselves observe observers, this property of intentionality is unavoidable.

Keeping in line with physics, should we declare an embargo on all the psychological traits we talk about in everyday life: consciousness, thought, beliefs, desires? Should we adopt the elaborate sanitary regimes of behaviorism? Should amorous partners say to each other: "That was good for you; was it good for me too?" The ludicrousness of this last resort becomes evident when we consider the denial it entails. Either we deny the existence of what we experience before we "become scientists" (for example, our own awareness), or we declare that science (read "physical science") cannot deal with such matters.

It is here that the second great figure of the scientific revolution of the seventeenth century, Descartes, comes to the fore. In his search for a method of thought, he was led to declare for "substance dualism." As I mentioned earlier, according to this view the world consisted of *res extensa* (extended things) and *res cogitans* (thinking things). Galilean manipulations work on *res extensa*, the set of extended things. But *res cogitans*, the set of thinking things, does not exist properly in time and space; lacking location, not being an extended thing, it cannot fall into the purview of an external observer. Worse still is the problem of interactionism: the mind and the body must communicate. With an uncharacteristic lack of clarity, Descartes declared that the pineal gland (figure 2–2) was the place where interactions between *res cogitans* and *res extensa* occurred.

Dualism has persisted in various forms to the present day. For example, while apparently monistic, behaviorism is simply dualism reduced by denial

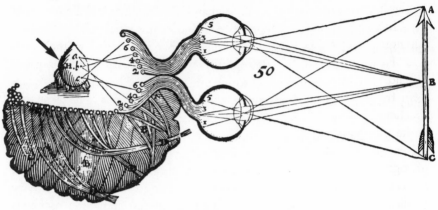

FIGURE 2–2

Diagram of the visual system from Descartes' Traité de l'Homme. *Descartes proposed that the retinal image from each eye was projected by nerve fibers onto the walls of the fluid-filled ventricles of the brain. It was assumed that binocular vision was then projected to the pineal body (black arrow). Descartes proposed this unpaired structure as the site at which* res cogitans *interacted with* res extensa—*soul with body. We now know that real binocular projection is to the visual cortex at the back of the brain on both the right and left sides.*

of the mind as a scientific object, and therefore left with one end hanging. Behaviorists solve the dilemma by examining behavior and ignoring intentionality. They do not attempt to put the mind back into nature; they simply deny its validity as a scientific object. And many nonbehavioristic psychologists, while asserting that they are materialists and not substance dualists, are nonetheless property dualists. While conceding that the mind and the brain arose from a single substance, they insist that psychological properties must be dealt with exclusively in their own terms, which necessarily differ from those used for the physical objects or bodies giving rise to these properties. A good example of a property dualist is Sigmund Freud in his later years.

I should point out that even remarkably accomplished biologists have been dubious about the enterprise of studying the mind. I once discussed these issues in a public symposium with the distinguished immunologist Sir Peter Medawar. He was just the slightest bit dismissive: "Of what use is it?" he asked. I managed to fend him off by pointing out that if we understood the brain better we could at least dispose of some crazy notions about how it works. Peter was the enemy of cant and my response quieted him.

I wish I could have said to him what Michael Faraday, I have heard, told the Chancellor of the Exchequer, William Gladstone, after presenting his findings on electricity. When the gentleman inquired loftily, "Of what use is it?" Faraday replied, "Sir, someday you shall tax it." (On another occasion he said, "Of what use is a newborn baby?")

In trying to put the mind back into nature, can we do better than substance dualism or property dualism? Or will we fall into further errors in the attempt? My answer to both questions is a qualified one. We can do better, but we cannot do it by assuming, as have some modern students of cognition, that the structure and biology of the brain are incidental and not central to the enterprise. Let us explore this issue further, for it has rich implications.

In the last few decades, practitioners in the field of cognitive science have made serious and extensive attempts to transcend the limitations of behaviorism. Cognitive science is an interdisciplinary effort drawing on psychology, computer science and artificial intelligence, aspects of neurobiology and linguistics, and philosophy. Emboldened by an apparent convergence of interests, some scientists in these fields have chosen not to reject mental functions out of hand as the behaviorists did. Instead, they have relied on the concept of mental representations and on a set of assumptions collectively called the functionalist position. From this viewpoint, people behave according to knowledge made up of symbolic mental representations. Cognition consists of the manipulation of these symbols. Psychological phenomena are described in terms of functional processes. The efficacy of such processes resides in the possibility of interpreting items as symbols in an abstract and well-defined way, according to a set of unequivocal rules. Such a set of rules constitutes what is known as a syntax.

The exercise of these syntactical rules is a form of computation. (At this point, please bear with me on what computation is exactly. For now, let us take it to mean the manipulation of symbols according to a definite procedure. I discuss this in detail in the Postscript.) Computation is assumed to be largely independent of the structure and the mode of development of the nervous system, just as a piece of computer software can run on different machines with different architectures and is thus "independent" of them. A related idea is the notion that the brain (or more correctly, the mind) is like a computer and the world is like a piece of computer tape, and that for the most part the world is so ordered that signals received can be "read" in terms of logical thought.

Such well-defined functional processes, it is said, constitute semantic representations, by which it is meant that they unequivocally specify what their symbols represent in the world. In its strongest form, this view proposes that the substrate of all mental activity is in fact a language of thought—a language that has been called "mentalese" (see the Critical Postscript).

This point of view—called cognitivism by some—has had a great vogue and has prompted a burst of psychological work of great interest and value. Accompanying it have been a set of remarkable ideas. One is that human beings are born with a language acquisition device containing the rules for syntax and constituting a universal grammar. Another is the idea, called objectivism, that an unequivocal description of reality can be given by science (most ideally by physics). This description helps justify the relations between syntactical processes or rules and things or events— the relations that constitute semantic representations. Yet another idea is that the brain orders objects in the "real" world according to classical categories, which are categories defined by sets of singly necessary and jointly sufficient conditions.

I cannot overemphasize the degree to which these ideas or their variants pervade modern science. They are global and endemic. But I must also add that the cognitivist enterprise rests on a set of unexamined assumptions. One of its most curious deficiencies is that it makes only marginal reference to the biological foundations that underlie the mechanisms it purports to explain. The result is a scientific deviation as great as that of the behaviorism it has attempted to supplant. The critical errors underlying this deviation are as unperceived by most cognitive scientists as relativity was before Einstein and heliocentrism was before Copernicus.

What is it these scholars are missing, and why is it critical? They are missing the idea that a description of the mind cannot proceed "liberally"—that is, in the absence of a detailed biological description of the brain. They are disregarding a large body of evidence that undermines the view that the brain is a kind of computer. They are ignoring evidence showing that the way in which the categorization of objects and events occurs in animals and in humans does not at all resemble logic or computation. And they are confusing the formal power of physics as created by human observers with the presumption that the ideas of physics can deal with biological systems that have evolved in historical ways.

I claim that the entire structure on which the cognitivist enterprise is based is incoherent and not borne out by the facts. I do not attempt to support this strong claim in the text of this book; to do so would require ranging over many disciplines with many unshared assumptions before arriving at my main thesis. For this reason, I have put my arguments against the forms of pure cognitivism into a Critical Postscript placed at the end of this book. Specialists may consult this essay at their leisure;

interested readers might more profitably turn to it after finishing the text of the book.

This essay addresses what I believe to be a series of category mistakes. The first is the proposal that the solution to problems of consciousness will come from the resolution of some dilemmas in physics. The second is the suggestion that computation and artificial intelligence will yield the answers. Third, and most egregious, is the notion that the whole enterprise can proceed by studying behavior, mental performance and competence, and language under the assumptions of functionalism without first understanding the underlying biology.

I will address the critical arguments in the Postscript. In the chapter that follows, I review some of the facts and ideas of biology and neuroscience. It is vital to understand the actual matter underlying the mind, and in particular its principles of organization. Only with such understanding will it be possible to dissect the difficulties we face when we attempt to study the mind, and to propose some ways out of the predicaments I have mentioned.

The principle I will follow is this: There must be ways to put the mind back into nature that are concordant with how it got there in the first place. These ways must heed what we have learned from the theory of evolution. In the course of evolution, bodies came to have minds. But it is not enough to say that the mind is embodied; one must say how. To do that we have to take a look at the brain and the nervous system and at the structural and functional problems they present.

CHAPTER 3

The Matter of the Mind

*The only laws of matter are those which our minds must fabricate,
and the only laws of mind are fabricated for it by matter.*
—James Clerk Maxwell

G iven the unique character of consciousness and the inability of
thought to "see into" its own mechanisms, it is no surprise that
some philosophers have proposed the idea of a thinking substance,
or even a kind of panpsychism in which all matter shares in consciousness. The results of modern investigations suggest, however, that the
physical matter underlying the mind is not at all special. It is quite ordinary—that is, it is made up of the chemical elements carbon, hydrogen,
oxygen, nitrogen, sulphur, and phosphorus, along with a few trace metals.
So there is nothing in the brain's essential composition that can give us a
clue to the nature of mental properties.

What is special is how it is organized. Those ordinary chemical elements
form parts of extraordinarily intricate molecules, which in turn make up
complex structures in the cells of living tissues. In a complex organism like
a human being, the cells come in about 200 different basic types. One of
the most specialized and exotic of these is the nerve cell, or neuron. The
neuron is unusual in three respects: its varied shape, its electrical and
chemical function, and its connectivity, that is, how it links up with other
neurons in networks.

I plan to tell you more about some of these properties, but only just enough to convince you that we are dealing with something unlike anything else in the universe. I shall add to this description as needed and in this way we won't be loaded down with intricacies right from the start. But before I lay out some descriptive details, it will be useful to give you a feeling for the numbers of neurons in certain brain areas and for the numbers of connections they make with each other. This will be, I think, a startling enough beginning. But when I get down to describing a bit of the morphology, you may be even more impressed with what evolution has accomplished in selecting for animals with richly structured brains.

Let us begin with the part of the brain called the cerebral cortex (figure 3–1), a structure that is central to what are loosely called the higher brain functions—speech, thought, complex movement patterns, music. If one were to take this corrugated "mantle" that covers the dome and the sides of your brain and spread it out, it would be the size of a large table napkin and about as thick. Counts of the nerve cells making up this structure are not very accurate, but it appears that there are about ten billion neurons in the cortex. (There are also other cells called glia that have supporting functions, but I will ignore them.)

Each nerve cell receives connections from other nerve cells at sites called synapses. But here is an astonishing fact—there are about one million billion connections in the cortical sheet. If you were to count them, one connection (or synapse) per second, you would finish counting some thirty-two million years after you began. Another way of getting a feeling for the numbers of connections in this extraordinary structure is to consider that a large match head's worth of your brain contains about a billion connections. Notice that I only mentioned counting connections. If we consider how connections might be variously combined, the number would be hyperastronomical—on the order of ten followed by millions of zeros. (There are about ten followed by eighty zeros' worth of positively charged particles in the whole known universe!)

So here we have our first clue as to what makes the brain so special that we could reasonably expect it to give rise to mental properties. And while the sheer number and density of neuronal networks in the brain are amazing, these are not the only unique properties of brain tissue. An even more remarkable property is the way in which brain cells are arranged in functioning patterns. When this exquisite arrangement of cells (their microanatomy, or morphology) is taken together with the number of cells in an object the size of your brain, and when one considers the chemical reactions going on inside, one is talking about the most complicated material object in the known universe.

I want to say a bit more about some properties of the brain's other

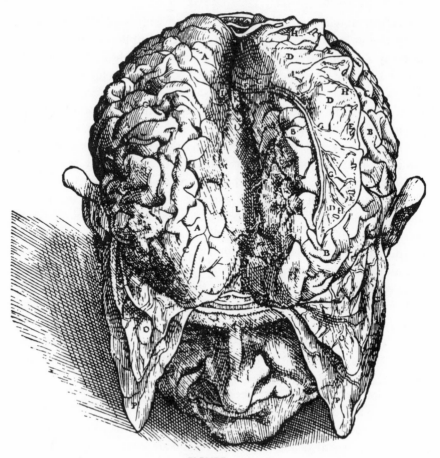

FIGURE 3–1

The exposed surface of the human cerebral cortex, drawn by the great anatomist Andreas Vesalius (1514–1564). He is generally considered to be the founder of modern anatomy, and his De Fabrica Humanis Corpora *set a new standard for medical art.*

components. In complex animals such as human beings, the brain consists of sheets, or laminae, and of more or less rounded structures called nuclei. Each of these structures has evolved to carry out functions in a complex network of connections, and each consists of very large numbers of neurons, sometimes more and sometimes less than in the cortex. The brain is connected to the world outside by means of specialized neurons called sensory transducers that make up the sense organs and provide input to the brain. The brain's output is by means of neurons connected to muscles and glands. In addition, parts of the brain (indeed, the major portion of its tissues) receive input only from other parts of the brain, and they give outputs to other parts without intervention from the outside world. The brain might be said to be in touch more with itself than with anything else.

How do neurons connect with each other and how are they arranged within nuclei and laminae? As mentioned, the major means of connection is the synapse, a specialized structure in which electrical activity passed down the axon of the presynaptic neuron (figure 3–2) leads to the release of a chemical (called a neurotransmitter) that in turn induces electrical activity in the postsynaptic neuron. As is suggested in the figure, the strength or efficacy of synapses can be changed—presynaptically by changes in the amount and the delivery of transmitter, and postsynaptically by the alteration of the chemical state of receptors and ion channels, the units on the postsynaptic side that bind transmitters and let ions carrying electrical charge (such as calcium ions) through to the inside of the cell.

Neurons come in a variety of shapes, and the shape determines in part how a neuron links up with others to form the neuroanatomy of a given brain area. Neurons can be anatomically arranged in many ways and are sometimes disposed into maps. Mapping is an important principle in complex brains. Maps relate points on the two-dimensional receptor sheets of the body (such as the skin or the retina of the eye) to corresponding points on the sheets making up the brain. Receptor sheets (for example, touch cells on your fingertips and retinal cells that respond to light) are able to react to the three-dimensional world and provide the brain with spatial signals about pressure or wavelength differences (they react to a four-dimensional world if we consider time as well). Furthermore, maps of the brain connect with each other via fibers that are the most numerous of all those in the brain. For example, the corpus callosum, the main fiber bundle connecting parts of your right brain to parts of your left across the midline, contains about 200 million fibers.

None of this was known in any detail before the nineteenth century. But remarkable surmises were made before that time by remarkable men— Denis Diderot, for example. The following is an excerpt from his novel in the form of a play, *Le Rêve de d'Alembert*, in which d'Alembert's mistress, Mademoiselle de l'Espinasse, queries the physician, Dr. Bordeu, about the causes of d'Alembert's disturbed dreams.

BORDEU: Because it is a very different thing to have something wrong with the nerve-centre from having it just in one of the nerves. The head can command the feet, but not the feet the head. The centre can command one of the threads, but not the thread the centre.

MADEMOISELLE DE L'ESPINASSE: And what is the difference, please? Why don't I think everywhere? It's a question I should have thought of earlier.

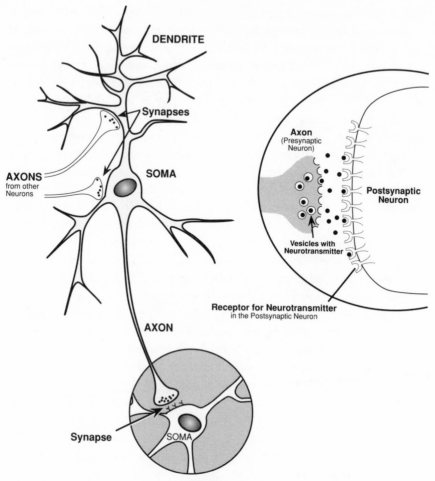

FIGURE 3–2

Some arrangements of the matter of the mind as exemplified by simplified neurons. Axons from near or distant neurons are long extensions that make contact with a neuron either on its body (soma) or on its branching processes, called dendrites. Axons carry electrical activity that causes the release of neurotransmitter when the electrical activity reaches the synapse with another neuron. After interacting with the appropriate receptors, the neurotransmitter in turn triggers the recipient (or postsynaptic) neuron to fire electrically. A simplified synapse is encircled at the bottom of this figure and enlarged in the circle at the right. The small round sacs represent vesicles in which neurotransmitter is stored in the presynaptic neuron. The postsynaptic neuron projects receptors (the Y-shaped structures) into the cleft between the presynaptic and postsynaptic membranes. These receptors bind to the transmitter released from the presynaptic vesicles and trigger the responses of the postsynaptic neuron.

BORDEU: Because there is only one centre of consciousness.

MADEMOISELLE DE L'ESPINASSE: That's very easy to say.

BORDEU: It can only be at one place, at the common centre of all the sensations, where memory resides and comparisons are made. Each individual thread is only capable of registering a certain number of impressions, that is to say sensations one after the other, isolated and not remembered. But the centre is sensitive to all of them; it is the register, it keeps them in mind or holds a sustained impression, and any animal is bound, from its embryonic stage, to relate itself to this centre, attach its whole life to it, exist in it.

MADEMOISELLE DE L'ESPINASSE: Supposing my finger could remember.

BORDEU: Then your finger would be capable of thought.

MADEMOISELLE DE L'ESPINASSE: Well, what exactly is memory?

BORDEU: The property of the centre, the specific sense of the centre of the network, as sight is the property of the eye, and it is no more surprising that memory is not in the eye than that sight is not in the ear.

MADEMOISELLE DE L'ESPINASSE: Doctor, you are dodging my questions instead of answering them.

BORDEU: No, I'm not dodging anything. I'm telling you what I know, and I would be able to tell you more about it if I knew as much about the organization of the centre of the network as I do about the threads, and if I had found it as easy to observe. But if I am not very strong on specific details I am good on general manifestations.

MADEMOISELLE DE L'ESPINASSE: And what might these be?

BORDEU: Reason, judgement, imagination, madness, imbecility, ferocity, instinct. . . .

BORDEU: And then there is force of habit which can get the better of people, such as the old man who still runs after women, or Voltaire still turning out tragedies.

(Here the doctor fell into a reverie, and MADEMOISELLE DE L'ESPINASSE said:) Doctor, you are dreaming.

BORDEU: Yes I was.

MADEMOISELLE DE L'ESPINASSE: What about?

BORDEU: Voltaire.

MADEMOISELLE DE L'ESPINASSE: What about him?

BORDEU: I was thinking of the way great men are made.

Even with the discussion so far, we are already in possession of a number of facts about the matter of the mind. Nerve cells are specialized, numerous, and hyperdense in their connections, which themselves have special chemical and morphological characteristics. The anatomy resulting from these arrangements is staggering in its intricacy and diversity. But it also has general organizing principles: It is made up of sheets that have topographic maps and of rounded nuclei, or "blobs." It sends multiple fibers to connect the maps to sensory sheets and out to the muscles of the body. And maps map to each other.

As a result of the stimulation of sensory elements, nerve signals in the form of electrical discharges occur at the membranes of neurons. They are caused by the flow of charged ions. (This means that electrical charges move more slowly in cell membranes than they do in telephone wires, where the current is carried by electrons.) Massive numbers of neurons act in parallel in amazing numbers of combinations. Their sensitivity to stimulation can be altered by a host of different chemicals, including the neurotransmitters at synapses, other substances called neuromodulators, and, of course, by drugs.

A piece of brain tissue is an intricate network that responds to electrical and chemical signals in three-dimensional space and in time. It sends out dynamic patterns and receives and responds to such patterns. These patterns affect each other and, through other nerve connections, the action of other organs of the body—the heart, kidneys, lungs, muscles, and glands. The brain is a master controller and its rhythmic patterns alter how you breathe, pump blood, digest your food, and move.

I discuss in a later chapter the principles by which the nervous system (and indeed the whole animal) develops both its overall and its microscopic shape. But it is useful to anticipate a bit of that discussion here. The anatomical arrangements of the brain and the nervous system are brought about by a series of developmental events (figure 3–3). In the embryo, cells divide, migrate, die, stick to each other, send out processes, and form synapses (and retract them). This dynamic series of events depends quite sensitively on place (which other cells are around), time (when one event occurs in relation to another), and correlated activity (whether cells fire together or change together chemically over a period of time).

Place dependencies in development are quite striking. Nowhere is this seen as clearly as in the formation of maps during embryonic development, such as the map of visual space formed by the retinotectal projection (figure 3–4). In this instance, neurites (fibers) from the ganglion cells of the retina form the optic nerve, the neurites of which then map in a definite fashion to a region that, in an animal like the frog, is called the optic tectum. Stimulation of a particular point on the retina by a point of light leads to the stimulation of neurons in a particular region of the tectum, and the responding cells are arranged in a definite map. Place is critical to the workings of such a map.

The arrangement of this map is achieved in at least two steps during development. The first step, which involves the extension of neurites in

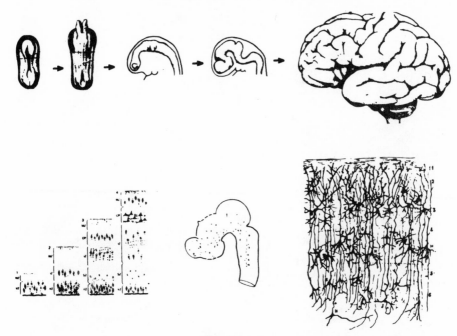

FIGURE 3–3

The development of the brain, from the neural groove (top left) to the cerebral cortex (top right). Individual neurons occur in layers (bottom left) or move in paths (bottom center), finally interacting in synapses to form an enormously complex neuroanatomy (bottom right). At one or another time in their careers all neurons are gypsies—moving to their final positions on other cells. The result is the most complicated material object in the known universe.

overlapping arbors by optic nerve fibers, forms a coarse map and does not require neural activity. The second step, in which the map is refined and becomes much more precise, requires neural activity in neighboring ganglion cell fibers that is correlated with neural activity in the tectum. Map formation during development in animals like the goldfish or the frog is dynamic, and the connections shift and reassemble as differential growth occurs both in the retina and the tectum. The principles governing these changes are epigenetic—meaning that key events occur only if certain previous events have taken place. An important consequence is that the connections among the cells are therefore not precisely prespecified in the genes of the animal.

What makes maps so interesting is that the epigenetic events that create form from place early in embryonic development must to some extent "anticipate" future interactions of the two-dimensional surfaces of sensory

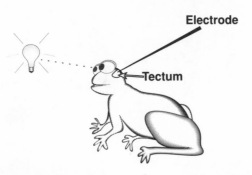

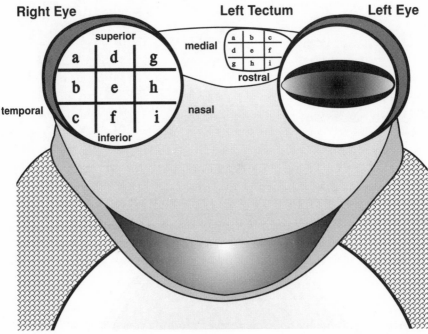

FIGURE 3–4

Mapping the eye and its visual fields to the brain. Top: *A frog with an electrode in the visual part of its brain (called the tectum).* Bottom: *Front view. The parts of the retina of the right eye marked with letters were each illuminated with light at the same time that an electrode measured electrical responses in the frog's tectum. Corresponding neighboring regions in the left tectum form a map of responses that are marked by the same letters. Note that there is a rotation of the map but that neighbors remain the same in the eye and in the tectum on the opposite side. Temporal regions of the eye map to medial regions of the tectum, inferior regions map to rostral regions, and so forth.*

receptor sheets (for example, the retina or skin) with the three-dimensional world in which the animal moves and receives stimuli. We will see later how the principles of development and evolution account for these phenomena.

What I have described so far may sound like the organization of a vast telephone exchange or perhaps even that of a digital computer. In some ways, the brain does indeed behave like these systems. When we look in detail at the structural features and functional properties of the nervous system, however, the analogy breaks down, and we are confronted with a series of problems. The implications of these problems amount to a series of interpretive crises for neuroscience as well as for those sciences that depend on its conclusions.

The structural crises, which I described in detail in my book *Neural Darwinism*, are those of anatomy and development. Although the brain at one scale looks like a vast electrical network, at its most microscopic scale it is not connected or arranged like any other natural or man-made network. As we have just seen, the network of the brain is created by cellular movement during development and by the extension and connection of increasing numbers of neurons. The brain is an example of a self-organizing system. An examination of this system during its development and of its most microscopic ramifications after development indicates that precise point-to-point wiring (like that in an electronic device) cannot occur. The variation is too great.

Furthermore, although the connectivity of neuronal systems in the central nervous system (particularly those that are mapped) is more or less similar from individual to individual, it is not identical. Indeed, as figure 3–5 shows, there is considerable variation both in the shapes of individual neurons in a class and in their connection patterns. This is not surprising, given the stochastic (or statistically varying) nature of the developmental driving forces provided by cellular processes such as cell division, movement, and death; in some regions of the developing nervous system up to 70 percent of the neurons die before the structure of that region is completed! In general, therefore, uniquely specified connections cannot exist. If one were to number the branches of one neuron and to number in a corresponding manner the neurons it touched, the numbers would not correspond exactly in any two individuals of a species—not even in identical twins or in genetically identical animals.

To make matters even more complicated, neurons generally send branches of their axons out in diverging arbors that *overlap* with those of other neurons, and the same is true of processes called dendrites on recipient neurons (see figure 3–2). I gave an example of this when I

STRUCTURAL VARIABILITY

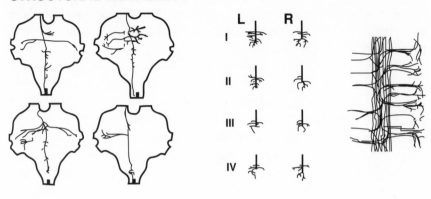

DYNAMIC VARIABILITY

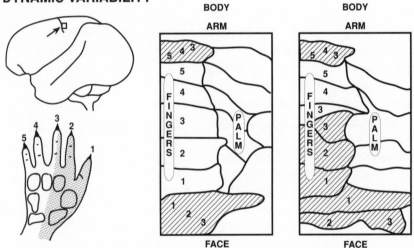

FIGURE 3–5

The variability of neural patterns. Top left: Patterns of the same nerve in four different locusts (the work of Keir Pearson and Corey Goodman). Top center: Visual neurons from four different but genetically identical water fleas (visualized by Edwardo Macagno and his colleagues). Note that even corresponding neurons on the right and left sides of one individual are not the same. Top right: Repeating nerve structures from a rabbit brain (see right-hand side) are all different. Bottom: The dynamism and variability of brain maps for touch in an adult owl monkey. A normal map (arrow indicates location in brain) of the fingers and palm (light numbered areas) and corresponding regions of the back or hairy side of the hand (dark areas) are shown in the map in the middle. After cutting the nerve that serves part of the front (hairless) side of the fingers and the palm, a rearrangement occurs in the map borders, not only of the front and the back, but also of other fingers (map at right). (This is the work of Michael Merzenich and his colleagues.)

discussed the arborization of optic nerve fibers on the tectum. To put it figuratively, if we "asked" a neuron which input came from which other neuron contributing to the overlapping set of its dendritic connections, it could not "know."

The existence of developmental principles leading to variance in connections and to overlapping arbors with unidentifiable (and not necessarily repeatable) patterns of synapses creates a crisis for those who believe that the nervous system is precise and "hardwired" like a computer. We may ask, "How has this crisis been met, when it has been recognized at all, by those who believe in the idea of the brain as a computer?

First, these explanations dismiss variations below a certain microscopic level as "noise," a necessary consequence of the developmental dilemma. Second, they deal with the absence of uniquely specified connections by arguing that higher levels of organization such as maps either do not need such connections or compensate for their absence in some fashion. And third, they explain the absence of precisely identified synaptic inputs by assuming that neurons use a code similar to those used to identify phone credit card or computer users. In neurons, the place and time codes presumably relate to the frequency, spacing, or type of neuronal electrical activity, or to the kinds of chemical transmitters with which they are associated (figure 3–2). Notice, however, that these explanations assume that individual neurons carry *information,* just as some electronic devices carry information. I argue later that this is not a defensible assumption and that these explanations are inadequate. No convincing evidence for the kinds of codes that humans use in telegraphy, computing, or other forms of human communication has been found in the human nervous system.

This brings us to some deeper riddles for those who would propose that the brain is a kind of computer. These riddles constitute a set of functional crises pertaining to physiology and to psychology. The first is this: If one explores the microscopic network of synapses with electrodes to detect the results of electrical firing, the majority of synapses are not expressed, that is, they show no detectable firing activity. They are what have been called "silent synapses." But why are they silent, and how does their silence relate to the signals, codes, or messages that they are supposed to be carrying?

A second dilemma concerns the functions and interactions of maps of the kind we have already considered for the retinotectal system. Despite the conventional wisdom of anatomy books, these maps are not fixed; in some brain areas, there are major fluctuations in the borders of maps over time. Moreover, maps in each different individual appear to be unique. Most strikingly, the variability of maps in adult animals depends on the available signal input (figure 3–5). This might not seem to pose a dilemma at first; after all, computers change their "maps" or tables on the alteration of software. But the *functioning* maps of the nervous system are based on *anatomical* maps—and at this anatomical level, they are changed in the adult brain only by the death of neurons. If the functioning neural maps are changing as a result of "software" changes, what is the code that gives two different individuals with variant anatomical maps the same output or result? One standard explanation is to say that there are alternative systems in the brain that handle changing input, each alternative fixed and hard-wired but switched in or out by changing input. The facts show, however, that the variance of neural maps is not discrete or two-valued but rather continuous, fine-grained, and extensive. Thus, the number of alternatives would have to be very large.

Another set of observations brings us to psychological dilemmas of the most profound kind. They cast doubt on the idea that the complex behavior of animals with complex brains can be explained solely by "learning." Indeed, this crisis highlights the fundamental problem of neuroscience: How can an animal initially confront a small number of "events" or "objects" and after this exposure adaptively categorize or recognize an indefinite number of novel objects (even in a variety of contexts) as being similar or identical to the small set that it first encountered? How can an animal, in the absence of a teacher, recognize an object at all? How can it then generalize and "construct a universal" in the absence of that object or even in its presence? This kind of generalization occurs without language in animals such as pigeons, as I discuss later on.

Explanations of these challenging problems tend either to rely on the existence of hidden cues, not obvious to the experimenter, or to treat the world of the responding organism as if its "objects" or "events" came with labels on them. But in reality, the world, with its "objects," is an unlabeled place; the number of ways in which macroscopic boundaries in an animal's environment can be partitioned by that animal into objects is very large, if not infinite. Any assignment of boundaries made by an animal is relative, not absolute, and depends on its adaptive or intended needs.

What is striking is that the ability to partition "objects" and their arrangements depends on the functioning of the maps that we discussed earlier. But how do maps interact to give definition of objects and clear-cut action or behavior? In human beings, a consideration of this question leads to what I call the homunculus crisis: the unitary appearance to a perceiver of perceptual processes that are known to be based on multiple and complex parallel subprocesses and on *many* maps. (In the visual system, there may be more than thirty interconnected brain centers, each with its own map.) Who or what organizes a unitary picture? "Computations" or "algorithms" in the brain—or the homunculus, a little man who has in his head yet another homunculus (see figure 8–2), and so on ad infinitum? Who is at home? If it is the homunculus, how could he have been constructed during the developmental wiring of the brain by his cousin, whom we may call the electrician? We have already seen that if he exists during development, this electrician has constructed some very odd wiring indeed.

Where does this leave us? The short answer is, "with a very great challenge." Unless we wish to pursue brain science in a purely empirical fashion without concern for coherent explanations, we have to confront the crises I have discussed here. An obvious alternative is to have a scientific theory that reconciles the apparent contradictions and dilemmas and resolves the crises. Clearly any satisfactory developmental theory of higher brain function must remove the need for homunculi and electricians at any level. At the same time, the theory must account for object definition and generalization made on a world whose events and "objects" are not prelabeled by any *a priori* scheme or top-down order. This sounds less like the tasks to which computers are put and more like something utterly unusual and very different from computers.

What is special about brains that computers, and material particles, and atoms, and *res cogitans*, and ghosts all lack is evolutionary morphology. As we have seen, this morphology interacts at many levels, from atoms up to muscles. The intricacy and numerosity of brain connections are extraordinary. The maps that "speak" back and forth are massively parallel and have statistical as well as precise features. Furthermore, the matter of the mind interacts with itself at all times. I have not yet mentioned that the dynamic arrangements of the brain show the system property of memory: previous changes alter successive changes in specified and special ways. Nervous system behavior is to some extent self-generated in loops; brain activity leads to movement, which leads to further sensation and perception and still further movement. The layers and the loops between them are the most intricate of any object we know, and they are dynamic; they continually change.

Indeed, the chemical and electrical dynamics of the brain resemble the sound and light patterns and the movement and growth patterns of a jungle more than they do the activities of an electric company. These dynamics result from a special chemistry. Alterations of that chemistry or destruction of its anatomical substrate can lead to temporary or permanent mental changes from elation to unconsciousness to death.

While we recognize that the marvelous matter underlying the mind is like no other, we must beware of a shallow chauvinism. Such a position would assert that *only* those biochemicals of which the brain is made could lead to such a structure. For even if that were to some extent the case, it is the dynamic arrangement of these substances to create mental processes, not their actual composition, that is essential. It is dynamic morphology all the way down. But not so far down that we have to invoke very special physical events and forces such as those between fundamental physical particles. Some scientists, ignorant of brain morphology and the properties of memory, have been tempted to explain mental properties at this level, the quantum level (see the Postscript).

If strict biochemical chauvinism is out, however, so is the liberalism of the computer scientist who assumes a brain software that actually does not exist *a priori* and then claims that it doesn't matter what structure this software runs on. He makes two fundamental errors, for there is no such thing as software involved in the operations of brains, and the evidence overwhelmingly indicates that the morphology of the brain matters overwhelmingly.

With this background and some of the neuroscientific facts in hand, we may now turn to more general biological matters. It is important to examine them if we are to avoid the pitfalls in our path toward a better understanding of the matter of the mind.

PART II

ORIGINS

One of the temptations of having a mind is to try using it alone to solve the mystery of its own nature. Philosophers have attempted this since time immemorial. Psychologists fall back on it, as do we all from time to time. But as a general method to explore the matter of the mind, it just won't do.

We have been in possession of an enormous insight into how our minds might work ever since Darwin proposed that minds arose by evolution. What this means is that minds have not always been around; they appeared at some definite time in a series of graded steps. It also means that we have to pay attention to animal form, because evolution teaches us that the selection of animals formed to carry out functions that increase their fitness is at the very heart of the matter.

At "the brain of the matter" is the most complicated arrangement in the known universe. To understand it will take us from philosophy to embryology, in a curious but necessary leap. When we have taken it, we will be in a position to return to philosophy via biology in the next two parts.

CHAPTER 4

Putting Psychology on a Biological Basis

Psychology was to him a new study, and a dark corner of educa-
tion. . . .
He put psychology under lock and key; he insisted on maintain-
ing his absolute standards; on aiming at ultimate Unity. The
mania for handling all sides of every question, looking into every
window, and opening every door, was, as Bluebeard judiciously
pointed out to his wives, fatal to their practical usefulness in
society. —Henry Adams

I gnoring the origins of things is always a risky matter. It is even more risky in any effort that purports to explain mental events. But this is exactly what has happened in much of the history of psychology and the philosophy of mind. I guess this is so because thought is a reflexive and a recursive process. It is therefore tempting to think that the nature of thinking can be uncovered by thinking alone. But if we go back to the earlier chapter on mind, we notice that the biggest difference between intentional objects and nonintentional objects is that the former are biological entities. The point is not that all living things are intentional, just that no nonliving things are. As I mentioned in the last chapter, we must account for how embodiment occurs in each individual.

So we must pay attention to biology. But embodiment is not the only reason for doing so. Equally important are the facts of evolution, which suggest that intentionality emerged rather late. What is the basis of the mental, and when did it emerge in evolutionary time? The glib answer is that the mental emerged when animals developed nervous systems. That is not quite correct, however; the mere possession of nerve cells does not appear sufficient. In this part of the book, I want to look at this question of origins. My goal is to demonstrate that the minimum condition for the mental is a specific kind of morphology.

Before I get to that, however, I want to make the more general case for linking psychology to biology. I shall do this in part by considering how philosophers proceeding in the absence of biology have been misled. I then want to show how too narrow a view of psychology can also lead us astray. In doing so, I do not wish to claim that the pursuits of philosophy and psychology independent of biology have been worthless. Often they *had* to be pursued in the absence of fundamental biological data. Even a wrong belief or a wrong theory can lend energy to a science, sustaining it until the appropriate evidence or methodology is available. So this chapter may be looked at as an historical interlude, shallow and brief, but revealing, I hope, of the rich skein of thoughts that have been brought to the matter of the mind.

The practice of ignoring biology when thinking about the mind and about how knowledge is acquired, without making reference to biology, has a distinguished history. To a great extent, the philosophy of mind has pitched its inquiries without concerning itself (except anecdotally) with the body or the brain. We have already seen that the first modern philosopher, Descartes, based his form of rationalism on thought itself using his well-known "method of doubt," which he outlined in the *Discourse on the Method*:

> I thought that I must . . . reject as if it were absolutely false everything about which I could suppose there was the least doubt, in order to see if after that there remained anything which I believed which was entirely indubitable. So, on the grounds that our senses sometimes deceive us, I wanted to suppose that there was not anything corresponding to what they make us imagine. And, because some men make mistakes in reasoning—even with regard to the simplest matters in geometry—and fall into fallacies, I judged that I was as much subject to error as anyone else, and I rejected as unsound all the reasonings which I had hitherto taken for demonstrations. . . . I resolved to pretend that everything which had ever entered into my mind was no more veridical than the illusions of my dreams.

Descartes' conclusion that there was a thinking substance radically sidestepped biology, along with the rest of the materially based order. Given his remarkable forays into biology, this is surprising. One matter Descartes did not explicitly analyze, however, was that to be aware and able to guide his philosophical thought, he needed to have language. And for a person to have language, at least one other person must be involved, even if that person is the memory of someone in one's past, an interiorized interlocutor. This requirement shakes Descartes' notion that his conclusions depended on himself alone and not on other people. Moreover, Descartes was not explicit as to when a human being first has access to a thinking substance in his development. Perhaps he should have pondered further the likelihood of a French baby concluding, "Je pense donc je suis."

Philosophical answers to questions of Cartesian rationalism, such as those provided by the British empiricists John Locke, George Berkeley, and David Hume, do not fare much better. Locke's notion of the mind as an empty slate, or *tabula rasa*, was explored in the absence of knowledge about developmental or evolutionary events indicating that entire behavioral repertoires can be under genetic control. And Berkeley's monistic idealism—suggesting that inasmuch as all knowledge is gained through the senses, the whole world *is* a mental matter—falters before the facts of evolution. It would be very strange indeed if we mentally created an environment that then subjected us (mentally) to natural selection.

The most ruthless and skeptical of the empiricists, Hume, concluded that no knowledge could be secure given that it is all based on sense impressions. Even scientific knowledge appeared to be shaken by his analysis of cause and effect as no more than mental correlation based on the repetition of these sense impressions. But as we will see later, sense impressions are not the issue; the biology of mind involves much, much more.

Immanuel Kant (figure 4–1), whose background in physics and astronomy was greater than in biology, put the matter in larger perspective. He answered Hume by pointing out the existence of categories *a priori* in the mind, thus assuring their coexistence with sensory experience. But while the existence of *a priori* categories appears in better accord with modern evidence on ethologically determined action patterns and on the neurophysiological properties of brain cells, it is not strictly consistent with developmental studies of how babies gain a sense of space, or even with the physics of relativity. Ignorant as he had to be of modern developments in biology and physics, Kant is to be forgiven for not understanding what constraints there might be on the *a priori*.

I could give other examples, but these should suffice to indicate that, in philosophy, a knowledge of psychology based on experiment and an understanding of neurology and evolution are useful to guard against extreme errors. But all this knowledge is a recent acquisition, and one can only admire the courage and persistence of these great thinkers in keeping important questions alive.

FIGURE 4–1

Immanuel Kant (1724–1804), the great philosopher of the Enlightenment. His profound ideas reshaped rationalist and empiricist pictures of the mind. The cartoon on the facing page depicts the great man preparing mustard.

Psychology itself has not fared very well in the absence of knowledge of the brain and nervous system. This is not to say that an enormous amount of useful and important information has not been accumulated since William James at Harvard in 1878 and Wilhelm Wundt in Leipzig in 1879 founded the first laboratories of experimental physiological psychology. Instead of a unified theory of the mind, however, a series of schools subsequently sprang up, each with different views on behavior, consciousness, and on the relative significance of perception, memory, language, and thought.

This is no place to review any of these schools at length. But it may be useful to mention some of their main lines of thought to underscore the need for a biological common denominator. James himself was one of the greatest pioneers of modern psychology. In *Principles of Psychology*, he argued that, while paying attention to the brain, psychology could proceed on its own, investigating mental functions by whatever combination of introspection, experiment, and psychophysics proved most revealing. Psychophysics was also advanced by Wilhelm Wundt, Ewald Hering, and the great physicist Hermann von Helmholtz during the same era in Germany.

FIGURE 4–1 *(continued)*

It consisted of careful measurements of reaction times and of judgments in response to accurately measured physical stimuli.

James's greatest achievement may have been to point out that consciousness is a process and not a substance in his characterization of this elusive process in his essay "Does Consciousness Exist?", a question he also pursued in *Principles.* Whitehead has made the claim that, with this inquiry, James was to the twentieth century what Descartes was to the seventeenth. During James's time, however, excessive attempts were still being made to use introspection to reach conclusions about the mind, often with dubious results (as in the case of Edward Titchener, who regarded experimental introspection as the "sole gateway to psychology" and elaborated grand theories of sensation and feeling based on this method). Similarly, studies of human memory (for example, by Hermann Ebbinghaus) used abstract or nonsense sequences and syllables, while paying little or no attention to the role of meaning in memory.

Ivan Pavlov's early-twentieth-century experiments on conditioned reflexes offered a strong reaction to these approaches. Animals receiving an unconditioned stimulus (food) paired repeatedly with a conditioned stimu-

lus (bell) salivated when presented later with the bell alone. Edward Thorn-
dike and Clark Leonard Hull in the United States extended and deepened
the study of stimulus-response paradigms. Eventually, the extreme position
emerged that the only scientific study possible in psychology was the
study of behavior. As enunciated by John Watson, behaviorism left consciousness, introspective reports, and the like outside the pale. The fiercest
latter-day advocate of this position was B. F. Skinner, who extensively
explored the phenomenon of operant conditioning. (Instead of responding
to a classical conditioned stimulus, an animal is rewarded during a particular
behavior or operant. This behavior is then reinforced by repeated reward.)

Many refined chains of behavior were analyzed using behavioral techniques. Clearly, however, at least part of the baby went out with the
bathwater. For example, these approaches did not encompass the Gestalt
phenomena (figure 4–2) discovered by Max Wertheimer, Wolfgang
Köhler, and Kurt Koffka. Gestalt patterns were discerned by thinking
subjects in a way that behaviorism was hard put to explain. Consciousness
simply would not go away. And the observations of Sigmund Freud, who
noted the effects of repression on memory and of the unconscious on
conscious behavior, pointed up the deficiencies of the behaviorist account.
The experiments of Sir Frederic Bartlett on human memory indicated that
more was involved in memory than the rote repetition of meaningless
strings of characters, as the previous work of Ebbinghaus had seemed to
imply. Biology and human nature were making strong claims that behaviorism had ignored.

One important aspect of human nature and behavior that needed to be
accounted for was revealed in the medical clinic. The discovery of brain
maps in the nineteenth century by Gustav Fritsch and Julius Hitzig, who
noted specific bodily movements in patients after electrically stimulating
parts of their brains, and the discovery by Paul Broca that damage to a
specific part of the left brain led to motor aphasia (the inability to produce
coherent speech), could not be ignored. In short order, schools of neurophysiology developed, and by the turn of the century scientists were well
on the way to measuring actual neural activity. Between the two world
wars, a series of technical innovations developed by Sir Charles Sherrington made it possible to detect both the individual and collective activity
of nerve cells.

Thus the picture of psychology was a mixed one: behaviorism, gestalt
psychology, psychophysics, and memory studies in normal psychology;
studies of the neuroses by Freudian analysis; clinical studies of brain lesions
and motor and sensory defects; classification of the psychoses with their
baffling symptoms in medicine; and a growing knowledge both of neuro-

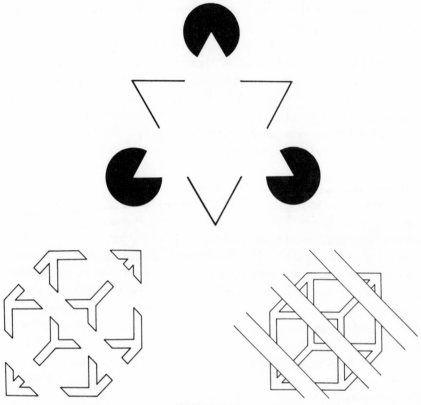

FIGURE 4–2

Gestalt phenomena. These figures are from Gaetano Kanizsa's work and show how context-dependent perception is. (Cover the Pac-Man figures on the top and watch the apparent contours disappear.) As Kanizsa put it, "Seeing and thinking are clearly distinguishable activities. With these 'pieces' we can imagine a cube (figure on the bottom left), but it is very difficult to see it." Notice, however, that the cube is completed (figure on the bottom right) behind the three opaque stripes and becomes perceptually present.

anatomy and of the electrical behavior of nerve cells in physiology, the first brought about by the neuroanatomical work of Santiago Ramón y Cajal and the second by the seminal physiological work of Sherrington. Only occasionally were serious efforts made by researchers such as Karl Lashley and Donald Hebb to connect these disparate areas in a general way. For the most part, each was pursued independently of the others, and research was sometimes accompanied by truculent denials of the applicability of competing ideas held by practitioners in "outside" fields.

What is curious about these developments is their relative separation from the theory of evolution, a theory absolutely essential to understanding the matter of the mind. Darwin enunciated the theory of natural selection in 1859. It was clear to him that evolution had to affect behavior and vice versa. But only Darwin's contemporaries George Romanes and C. Lloyd Morgan promulgated the idea that a connection existed between evolution and behavior. The effects of development on behavior were appreciated by C. W. Mills and J. M. Baldwin, but their insights, which are part of the foundation for our modern view, did not penetrate the mainstream. Later developmental studies by Jean Piaget of the cognitive behavior of children laid the groundwork for modern studies of cognition in development.

Of course there have been elaborations at the other extreme: efforts have been made to explain behavior in terms of social psychology ever since those of Herbert Spencer, another of Darwin's contemporaries. These efforts have generally been at a descriptive level and have invoked cultural traits and "folk psychology"—the commonsense evaluation of human behavior. Controversies and speculations about nature and nurture, genetics and environment, have abounded. This rich and sprawling set of studies does not lend itself easily to synthesis. Nonetheless, as modern methods of measuring brain function developed and an increased understanding of brain biochemistry emerged, it became clear that psychology could not be pursued without being increasingly grounded in biology. At best it could be provisionally pursued (as it always has been) while awaiting biological interpretation.

Once one arrives at this conclusion, however, there is no escaping an even more fundamental one: The phenomena of psychology depend on the species in which they are seen, and the properties of species depend on natural selection. This view, taken by ethologists such as Nikolaas Tinbergen and Konrad Lorenz and also by most modern psychologists, inexorably links psychology to biology. That linkage demonstrates the importance of evolutionary origins in the behavior of species.

In considering our minds, we must also consider both our kinship with and our differences from other species. As I discuss in chapter 16, one difference is that each of us has an individual "soul" based on language. Whatever we find out about the properties of language, however, the sad fact is that neither psychology nor biology will permit the transmigration of souls. The tale is told of a dying man who consoled his already grieving wife with a promise that he would return exactly six weeks after his demise, at which time she was to visit a medium. Comforted, she waited patiently and went to the medium on the appointed day. A voice from a dark corner of the room said, "Hello, darling." "Harry," she said, "is that you?" The voice said, "Of course it's me." Somewhat gingerly, she asked, "What do you do every day?" The voice replied, "I get up, make love, take a walk, make love, eat, make love, nap, make love. The next day, the same old thing." She took this in and said carefully, "But, darling, I didn't know the angels in heaven made love." The voice replied, "I'm not an angel in heaven, I'm a rabbit in Saskatchewan."

While the ideas of philosophers and of different psychological schools must be taken into account in any consideration of the matter of the mind, such ideas have only lately come to grips with the key issues of biology itself. The message boils down to this: The fundamental basis for all behavior and for the emergence of mind is animal and species morphology (anatomy) and how it functions. Natural selection acts on individuals as they compete within and between species. From studying the paleontological record it follows that what we call mind emerged only at particular times during evolution (and rather late at that).

These terse comments can be used as the basis for a research program to connect psychology with biology—a program to account for embodiment. Given the record of the history of the philosophy of mind and of psychology, the continued avoidance of the biological underpinnings of such a program is not likely to enhance our understanding of how the human mind emerged and how it functions. Errors continue to arise when psychology is pursued without strong connections to biology; I discuss some of them in the Postscript.

The center of any connection between psychology and biology rests, of course, with the facts of evolution. It was Darwin who first recognized that natural selection had to account even for the emergence of human consciousness. Let us turn to some of his insights and their consequences.

CHAPTER 5

Morphology and Mind:
Completing Darwin's Program

*But then arises the doubt: can the mind of man, which has,
as I fully believe, been developed from a mind as low as that possessed by the lowest animal, be trusted when it draws such grand
conclusions?*   —Charles Darwin

Alfred Wallace, the codiscoverer of the theory of natural selection,
wrote a series of letters to Charles Darwin expressing what he felt
was a heretical view. Wallace denied that natural selection could
account for the evolution of humans, arguing that the capabilities
of the human mind could not be explained by natural selection alone.

Darwin (figure 5–1) took the opposite position. He saw no reason why
natural selection could not have given rise to the basic features underlying
human thought. His books *The Descent of Man* and *The Expression of the
Emotions in Man and Animals* were dedicated to this idea.

It is important to understand Darwin's ideas on evolution and natural
selection. Simply put, they state that evolution occurs as a result of competition and environmental change, both of which act on variation in populations (figure 5–2). Variation always exists in living populations, and it
results in differences in fitness. Natural selection results in the differential
reproduction of those individuals whose variations (read "structural and
functional capabilities"—their phenotype) provide them and their progeny
with statistical advantages in adapting to environmental change or in
competing with individuals of the same or different species. Differential

FIGURE 5–1

Charles Darwin (1809–1882), the founder of modern evolutionary theory, the theoretical basis of all of biology. Darwin insisted that the evolution of humans as a species was subject to the same kinds of forces as those leading to the evolution of other species. He became retiring in his later life, and this dignified but somewhat lugubrious picture is indicative.

reproduction and heredity enhance the likelihood that the traits that increase fitness will be preserved.

What is critically changed in the resulting population is the frequency of the genes that give rise to those traits. (I discuss genes and genetics in more detail in the next chapter.) The *fact* that evolution has occurred is scored by the change in gene frequencies. But the *means* by which it occurs is natural selection on the phenotype (the total structural and functional capabilities) of individuals. The main *level* at which selection occurs is the

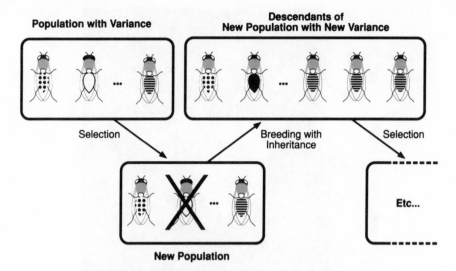

Population with Variance

**Descendants of
New Population with New Variance**

Selection

Breeding with
Inheritance

Selection

Etc...

New Population

POPULATION THINKING

FIGURE 5–2

Population thinking. Variation occurs by mutation in a population of organisms. Natural selection favors the differential reproduction of those members of the population that show greater fitness on the average. The result is that the relative frequency of genes conferring fitness increases in the population. Notice selection against dark eyes and for striped bodies.

individual and his behavior. What we need to understand are the rules connecting the ways in which genes are sorted and expressed with the ways in which genes lead to changes in the phenotype (figure 5–3). This is a formidable task, one that is only partly completed.

While Darwin did not grasp the correct genetic mechanisms, he got the principle right. He understood, for example, that phenotypic resemblances between the emotional expressions of certain animals and those of human beings were likely. He also understood that natural selection need not have selected all emotional expression *directly*. The same considerations apply to thought and behavior. His position was that gradual changes in populations could account even for the emergence and descent of human beings.

Now there was much in this position that could not be substantiated in Darwin's time. Many things were unknown to Darwin, including the true nature of genetic inheritance, essential data on hominid fossil remains, and a good deal of important information on how animals develop. But his basic approach, it turns out, was a sound one. It consisted of understanding what we need to know in order to understand the evolutionary origin of the human mind, what I call Darwin's program. What we need to under-

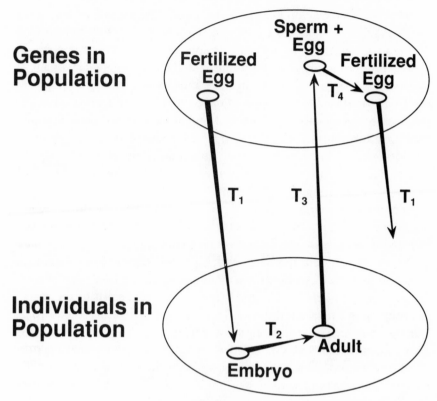

Genes in
Population

Individuals in
Population

FIGURE 5–3

Changes in the frequency of genes may be related to the actual process of natural selection. The result and indication of evolution is a relative increase in a population of those genes that have enhanced fitness. But selection occurs on individuals, whether on sperm or egg, or on the embryo, or on the animal in its environment. So we must understand the rules relating development and behavior to the effects of genes [the vertical lines for transformations $(T_1–T_4)$]. T_1 represents development; T_2 represents the conversion to adult life in the environment; T_3 represents the formation of sperm and eggs; T_4 represents the fertilized egg ready to undergo a new cycle of development.

stand (aside from the mechanisms of inheritance) is how the morphology underlying behavior arose during evolutionary history, and how behavior itself alters natural selection. I call this Darwin's program, not because it represents everything he wished to know, but because this was what most concerned him in his later years. Of course, Darwin did not complete his program. If we accept his position that there is no aspect of human behavior that cannot eventually be accounted for by an evolutionary explanation, then our task is to try to complete it.

What would be required to complete this program? First, an analysis of the effects of heredity on behavior, and vice versa. Second, an account of how behavior is influenced by natural selection and in turn influences it. Third, an account of how behavior is constrained and made possible by animal form or morphology. And fourth (and most fundamental of all), an understanding of how animal form arises and changes during development. (By form, I mean not only shape—limbs, symmetry, and so on—but also the microscopic details of tissues and organs such as the brain that give them their functions.) This last requirement entails an understanding of the relation between evolution and morphogenesis in development. The nature of this relation is the outstanding and central riddle of modern biology—that of morphologic evolution.

Here it may be useful to state what is known about the other requirements of Darwin's program. One is fairly complete: We know that the basis of heredity rests in the genes and we understand a good deal about how genes are transmitted, modified, and expressed (figure 5–4). It was the coming together of geneticists and evolutionists in the 1940s that allowed the genetic discoveries of Gregor Mendel to be connected to the theory of evolution in a most fruitful way. This "modern synthesis" accounted (as Darwin could not) for the origin of genetic variation as mutations in deoxyribonucleic acid (or DNA) as well as for the rearrangement of genetic structures in a process called recombination. In short, it began successfully to score the results of natural selection in terms of the changes of gene frequencies in populations.

Subsequent discoveries about the nature of DNA and the ability to manipulate this molecule, even to the point of inserting foreign genes into animals and changing their form or behavior, triumphantly confirmed the position taken in the modern synthesis. Moreover, great progress was made in extending Darwin's ideas about how different species arose through the sexual or geographic isolation of breeding groups of individuals.

Following the modern synthesis, a number of scientists began to study behavior in terms of genetics, evolution, and species interactions. This gave rise to the science of ethology, the data of which support the notion that some behavior patterns are species-specific and thus subject to genetic influence. The findings of ethologists are more subtle than this, however. They indicate that complex behaviors such as bird song have both genetic *and* epigenetic components. For example, some aspects of the motor patterns underlying the song of certain species such as song sparrows are given from birth as part of the phenotype. So are some variations and modifications of vocalization patterns. But to be able to sing the song characteristic of a song sparrow species in a given area, a sparrow needs to hear the songs of conspecifics—mature birds of the same species. In

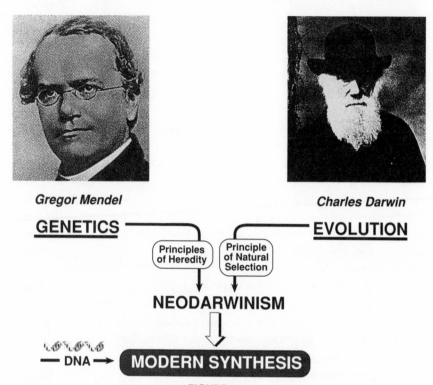

FIGURE 5–4

The modern synthesis. In the 1940s a group of evolutionists and geneticists reconciled Gregor Mendel's (1822–1884) original findings on heredity with the theory of evolution described by Darwin (1809–1882). Darwin's original theory had an incorrect notion of heredity. Mendel, an Augustinian monk from Austria, laid the foundations of modern genetics, but his contribution was not recognized at first and had to be rediscovered in 1901.

species like song sparrows, birds deafened from birth can never develop the full-fledged species-characteristic song. Epigenetic events involving interactions with other birds of the species are required for that.

It is not too difficult to see how patterns of behavior could affect and be affected by genetic variation and natural selection. Studies of this kind, which connect the functions of regions of the brain to the rest of the phenotype, go a long way toward filling in parts of Darwin's program. But one must not follow this approach to excess. Developments in the field of sociobiology may serve as a warning. Sociobiologists are concerned with how behaviors can be accounted for by natural selection. Altruism is a case in point. If natural selection occurs to maximize the fitness of *individuals*, it is difficult to see how the genes of individuals who sacrifice themselves before they breed, or who lose breeding potential in the service of others, could be passed on. The genetic analysis of bees undergoing what is called kin selection indicate that females serving a sister queen can, by forgoing their own reproduction, increase the frequency of their genes in a population. This finding, which depends on unusual features of the genetics of bees, is an elegant experimental triumph. But attempts to account for human altruism as a direct consequence of "genes for altruism" are another matter—and a dubious one at that.

Genes do not act directly, but rather in complex combinations, to alter form. And form alters behavior in subtle ways. More tellingly, subtle changes in form sometimes lead to rather extraordinary changes in behavior. What we want to know is how alterations in form, either in the whole animal or at microscopic levels of brain, muscle, or bone affect behavior, and how behavior alters form. This is the part of Darwin's program that remains largely incomplete.

One can appreciate how extraordinary a person Darwin was by poring over his journals (figure 5–5). In one of them, the M notebook, he says: "Origin of man now proved.—Metaphysic must flourish.—He who understands baboon will do more toward metaphysics than Locke." Sad in his last several decades, more or less reclusive, he steadfastly continued his indefatigable researches. Only now is it becoming clear how much he actually knew. He thought as deeply about behavior as he did about form. He stands as a profound thinker to whom it would not have occurred to attempt to appear "smart."

Incidentally, there is no escape from the difficulty posed by the idea that genes specify complex behavior directly by attempting to invoke "group selection." This is the notion, for example, that natural selection acts to favor quick herds of animals rather than by selecting quick individual animals that constitute a herd. Darwin raised and discussed such a possibility. With few exceptions, however, it appears that most natural selection occurs not at the level of genes or groups of individuals, but rather at the level of individuals themselves.

These considerations only emphasize again the part of Darwin's program that needs most to be completed. This part is concerned with how animal form, tissue structure, and tissue function could have arisen from ancestors—the problem of morphologic evolution. To see why this problem is so important, one need only think about the extraordinary evidence from fossil data on hominids indicating the large increase in hominid cranial capacity and brain size that has occurred over less than a million years of evolution (see figure 5–6).

How could this have occurred so rapidly? How does it relate to other

83e   The possibility of two quite separate trains going on in the mind as in double consciousness may really explain what habit is— In the *habitual* train of thought one idea. calls up other, & the consciousness of double individual is not awakened.— The habitual individual remembers things done in the other habitual state because it will (without direct consciousness?) change its habits.—

Aug. 16th. As instance of heredetary mind. I a Darwin & take after my Father in heraldic principle. & Eras a Wedgwood in many respects & some of Aunt Sarahs. cranks[1], & so is Catherine in some respects—. good instances.— when education same.— My handwriting same as Grandfather.[2]

84e   *Aug. 16th* Anger «Rage» in worst form is described by Spenser (Faery Queene. **CD 25** (Descript of Queen) «O» of Hell Cant IV or V.) as *pale & trembling*. & not as flushing & with muscles rigid.—[1] How is this? **dealt with p. 241**[2] Origin of man now proved.— Metaphysic must flourish.— He who understands baboon ‹will› would do more towards metaphysics than Locke

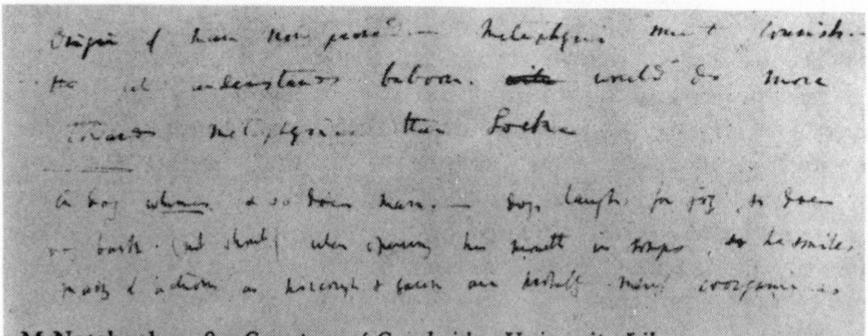

M Notebook, p. 84. *Courtesy of Cambridge University Library.*

FIGURE 5–5
An excerpt from Darwin's notebooks.

hominid traits we can guess about from the fossil record and from archeological remains? What is the connection between overall morphology and behavior and the microscopic morphology of the brain? How do these evolutionary developments connect with the behavior of hominids in groups and with the development of language?

These are profound and largely unanswered problems in paleontology, anthropology, and archeology. They are difficult because the record is fragmentary; the soft tissues are gone, leaving mainly bones, and thus structure and function can only be connected indirectly. But one thing is clear. Even if we had better evidence, we would still need a theory of how morphology arises and how it is changed during evolution.

Why is this so? Morphology—the shape of cells, tissues, organs, and finally the whole animal—is the largest single basis for behavior. There is much evidence to support this conclusion in a gross and even a trivial

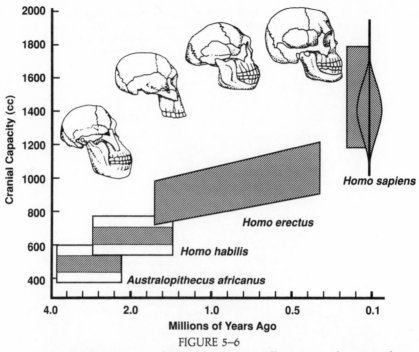

FIGURE 5–6

The remarkable increase in cranial capacity over two million years of human evolution. We are Homo sapiens sapiens; *the rest are our presumed ancestors. The shaded quadrilateral represents the time in which each species lived and its brain size. The* Homo sapiens *rectangle is for* Homo sapiens sapiens *and Neanderthal man. The dark spindle represents the range and cranial capacity in modern* Homo sapiens sapiens. *Originally, these skulls were not empty! An adequate brain theory must be able to account for such a large increase in brain size over such a short period of evolutionary time.*

sense. To fly you need wings; to think, a brain. But at another level, morphology is an extraordinarily subtle matter. The smallest change in the position of the insertion of a levator muscle in the jaws of cichlid fish allowed swallowing to occur independently of bait grasping. This provided the basis for an explosive increase in the occupation of different econiches by descendant variants of cichlid fish, an explosive adaptive radiation that outstripped many competitors. Morphology matters for us, too: If one compares humans and chimpanzees at the gene level, they show 99 percent identity to each other. But morphological change leading to the presence of sustained bipedalism, altered jaw muscle insertions into the skull, a larger cranium, a supralaryngeal space with speech organs, and a part of the cerebral cortex called the planum temporale appear to have been decisive in leading to characteristic human behavior.

There is evidence for a relation (but not a linear one) between the size and complexity of the brain and the complexity of behavior. There is much evidence for the coupling of functions of specific parts of the brain to specific skills. And clinical evidence on damaged brains indicates that specific, recognizable loss of functions of the mind occurs when particular brain regions are damaged. These various findings suggest that to understand the evolution of mind and behavior, we must first understand the bases of morphologic evolution.

Results like these support Darwin's supposition that human mental capacities arose by natural selection. There is diffuse evidence for mentation in the fossil and archeological records. Evidence of burial of the dead, for example, may be taken as evidence of human consciousness and possibly even of self-consciousness. But perhaps a more telling argument can be constructed by looking at how the human brain is structured, how it functions, how its cells arose, what they have in common with those of other species—and what is different and special about it.

This requires a theory of the evolution and development of animal and tissue form. It then requires that a theory of brain function be constructed based on the first theory, a tall order that must be filled if we are to complete Darwin's program. What makes the enterprise fundamentally interesting is that it is *not* specific to brains. To accomplish it, we need to show how development (embryology) is related to evolution. We need to know how genes affect form through development. We have to ask how this process constrains evolution—how the rules of development, which themselves evolved, can only be realized in particular ways.

This knowledge is necessary because evolution is historical, because only certain combinations of developmental events lead to functioning shapes, and because the shape of an animal's body is as important to the functioning and evolution of its brain as the shape and functioning of the brain are to the behavior of that body.

In the next chapter, we will look at how embryology and evolution *interact* to result in brains and bodies. A word of warning is in order: expect no miracles of simple explanation. Given what we know about evolution, it is no more likely that a gene can be found for altruism than that a single biochemical substance will be found to distinguish an ape from a human. The connections between morphology and mind that complete Darwin's program will be more indirect and circuitous than that. Yet their intricacy will make them all the more intriguing. Not the least of the intrigue is how a fertilized egg gives rise to a functioning animal, brain and all. This will, I hope, justify a change of pace to present a mini-course in modern molecular biology and development. We will need some of its lessons in the next part of this book, when we face the problem of constructing a brain theory that is consistent with evolution and development.

CHAPTER 6

Topobiology: Lessons from the Embryo

"The Chicken and the Egg, Together at Last"
—Title of a review of *Topobiology*,
New York Times Book Review, January 22, 1989

I t may seem strange that we must concern ourselves with embryology when this book is about the mind. Eggs and sperm show no evidence of mind, and neither do very early embryos. But since we know that newborn infants do show evidence of mind, however feebly, it seems reasonable to wonder by what interactions the bases for mental life have been laid down.

But why deviate to such issues as shape and form? And why concern ourselves with cells, molecules, and DNA? The straightforward answer is that the rules by which embryos are built govern the way that brains are built. The actual formation of the anatomy of the brain depends on muscles acting on bones, nerves acting on skin in a given order, and so on—that is, it depends on the rest of the phenotype. And as I stated in the last chapter, if we are to understand when aspects of mind arose in the course of evolution, we have to understand the connection between development and evolution.

To examine development, I will of necessity rely on some technical words and details. My suggestion to the reader is to go by the details once, look at the figures, and then return to the text. Let me dispose of some preliminaries (figure 6–1). The cells of higher organisms (called eukaryotes) have nuclei that contain DNA, the hereditary material. DNA is made up of long strings of four smaller molecules, called nucleotide bases, that are linked in sequences. There are only four types of bases (guanine, cytosine, adenine, and thymine, or G, C, A, and T for short). Thus, the sequence of a strand of DNA might look like this: . . . GTCGACCTGGCAGGT-CAACGGATC . . . It is now known that each strand of DNA containing such a string has a complementary strand coiled with it: . . . CAGCT-GGACGTCCAGTTGCCTAG . . . Complementary bases pair with each other: Notice that in this "double helix," G from one strand pairs with C from the other, and A with T, whereas *within* each strand the Gs, Ts, and so on are linked by strong chemical bonds, like beads on a string. By contrast, across strands the Gs pair with Cs and As with Ts by weak chemical forces that allow the two strands to come apart, for example, as the result of an increase in temperature.

The key points are these: 1. Using a single strand as a template, a second DNA strand can be built from single bases by special protein enzymes, which catalyze or speed up the chemical linkage of one base to the next in the strand being formed. The order in the new strand is determined by the pairing of the right base to its complementary partner on the opposite strand. 2. A sequence of three bases (any combination of G, C, A, and T) on a strand of DNA represents a code word (or codon) telling the cell to incorporate a particular protein building block called an amino acid into a long string of such amino acids, called a polypeptide. This polypeptide chain then folds up to form a protein. If each code word is three nucleotides long, sixty-four code words can be constructed from four kinds of nucleotides, yet only twenty amino acids occur in proteins. Obviously, then, some code words do not code for amino acids, while others are simply redundant. A piece of DNA of the right length and base sequence to specify a protein is known as a gene. 3. When a cell divides, it copies the DNA from one of the strands to provide new DNA for its two daughter cells. Normally, each copy will have exactly the same sequence of code words. If, however, a mistake is made or a DNA strand is cut, say by a cosmic ray, replicating or repair enzymes may not copy the template strand faithfully. This is one way in which mutational change is incorporated into a gene, altering its code.

The discovery that DNA was the genetic material was made at The Rockefeller Institute by Oswald Avery and his colleagues, and was reported in an extraordinarily significant paper in the *Journal of Experimental Medicine* in 1944. Curiously, the paper did not cause an immediate explosion of belief or interest. I used to play music with Stuart Elliot, a microbiologist who worked closely with Avery and also with Fred Griffith in England, another pioneer in this field. One day in 1964, Stuart suggested

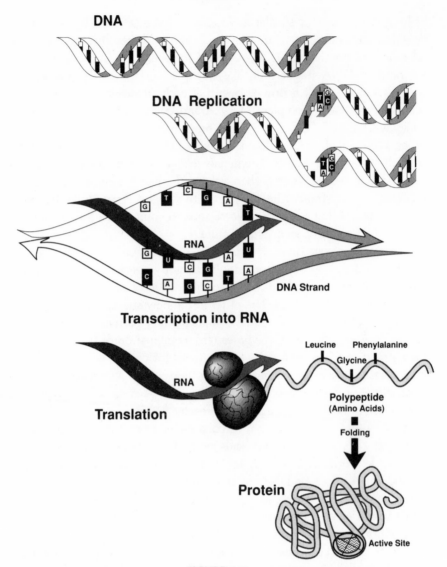

FIGURE 6–1

The reading of the genetic code into protein (a mini-course in molecular biology). DNA consists of two strands held together by weak forces between complementary nucleotide bases. Guanine (G) pairs with cytosine (C), while adenine (A) pairs with thymine (T). The bases within a single strand, linked by much stronger forces, can be read in sequence as a series of three-base code words, each specifying a particular amino acid (the building block of protein). While DNA itself stays in the nucleus of the cell, its code is carried elsewhere in the cell in the form of single-stranded RNA, which is built by special enzymes that transcribe the DNA sequence. (The RNA code uses one different base, uracil (U), which takes the place of thymine, but is otherwise the same as that of DNA; some typical code words are UUU = phenylalanine, CUU = leucine, GGC = glycine, and so that I approach Peyton Rous, the editor of the *Journal of Experimental Medicine*, to propose that the journal republish one of Griffith's original papers together with the great paper of Avery and his colleagues. This, Stuart suggested, would appropriately commemorate both the twenty-fifth anniversary of Griffith's death and the twentieth anniversary of the Avery paper.

Rous (who in his eighties won the Nobel prize for his early work on viruses that cause cancer) promised to "take up the suggestion with the board of the journal." After hearing nothing for six weeks, I encountered Rous stepping off a bus, and inquired. He stopped, stared gravely at me, and said, "Ah, yes. I took it up with the board and they thought it was an extremely vulgar suggestion." Surprised, I walked with him in silence for a block, and then I heard him say quietly, "I never did like Avery very much." When I asked him why, he said, "What would you think of a man who got a medal from the Royal Society and never went to pick it up?"

Avery did not get a Nobel Prize. His work was only fully appreciated some years after his death. Even now, with our molecular biological understanding of DNA, it is not easy to imagine how momentous his discovery was. Credit in science, as elsewhere, is not always evenly distributed.

This little précis of molecular biology allows us to say something already about the relation between the genotype (the set of genes possessed by an organism) and its phenotype. The genes consist of long strings of code words, with start and stop signals (these are also part of the code). Through the machinery of the cell, DNA is copied into another long string of somewhat different nucleotides called RNA. The RNA is then shipped out of the cell's nucleus to be read by a cellular device (like a tapehead) that brings amino acids corresponding to the coding sequence together to be linked in the proper order to make a polypeptide of perhaps several hundred amino acids in length (figure 6–1).

When finished, this polypeptide folds up in a complex shape to form a more or less compact protein (figure 6–2). The order of the amino acids in

forth.) This RNA code sequence is then "translated" on a special structure, the ribosome, where other enzymes link each amino acid in turn as the codes that specify them are read off the RNA. (The amino acids are carried to the site by another kind of RNA.) The polypeptide of linked amino acids that results folds into a three-dimensional shape that depends on its coded sequence. A single change, or mutation, in the DNA code may alter the sequence of amino acids, which may cause, in turn, the protein's shape to change. This can alter its function, which is carried out by an active site on the folded structure. When a cell divides, the two DNA strands separate and an enzyme copies each strand to give identical DNA to each of the cell's two daughter cells in a process called replication.

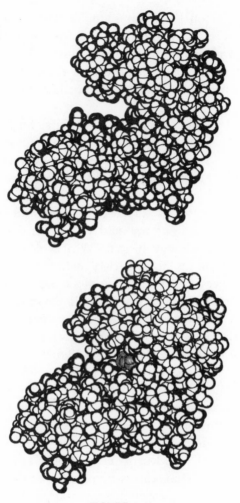

FIGURE 6–2

Protein folding and function—an example. A folded protein called hexokinase can bind the sugar glucose in the cleft that is the active site for its enzyme function, catalyzing a chemical reaction in which phosphorous derivatives are linked to glucose during metabolism. Top: *The protein in the absence of glucose.* Bottom: *When glucose is added the cleft closes around it, binding it securely.*

its chain determines the shape of the protein. The *shape* confers phenotypic properties and functions on a protein; for example, certain shapes might allow it to fit together (like blocks) with other proteins to form cell structures, while others might allow it to bind to chemicals and change the speed with which they react. As mentioned before, this is the key property of an enzyme. To summarize:

1. DNA "makes" RNA which "makes" protein (where the quotes mean "specifies"—it is the cell that actually makes the chemicals).
2. The shape of the protein depends on its sequence of amino acids, which depends in turn on the original sequence of the code words in the corresponding DNA.
3. The function of the protein depends on its shape.

Since much of an organism's phenotype depends on the properties of its proteins, these rules might seem to account not only for the shape of proteins but also, by extension, for the shape of animals.

Alas, things are not so simple, for it is not by building up *proteins* but by building up *cells* that an embryo is made. Its shape and that of its tissues, including the brain, derive from the shapes of collections of cells of a variety of types, each type with different proteins (the differences come from the fact that different combinations of genes are expressed in different types of cells).

So we must now ask how cells do this. But in doing so, we must not lose sight of a key point—the shape of the animal ultimately *does* depend on the order of the code words in its DNA. Moreover, changes in shapes over the course of evolution must have arisen from mutations that changed the order of the code words in the DNA of an ancestor. So the questions we want to answer are:

1. How does the one-dimensional genetic code specify the shape of a three-dimensional animal (not just a three-dimensional protein molecule)?
2. How can we account for changes over time in the developmental processes leading to such shapes so that new shapes evolve?

To answer these questions in a provisional way, I wrote a book called *Topobiology* explaining, among other things, how brains could have evolved. *Topos* means place, and the title refers to the fact that many of the transactions between one cell and another leading to shape are place-dependent: They occur only when a cell finds itself surrounded by other cells in a particular place. Let us consider some of the place-dependent events that lead to the formation of an embryo and its organs, particularly its brain.

An embryo is formed when a sperm containing DNA from a male fertilizes an egg containing DNA from a female. (By the way, the germ cells—sperm and egg—are enormously varied because each may contain genes with different mutations.) There are lots of genes and each has a long string of code words. The fused sperm and egg (or zygote, as it is called) now has genes from both parents and it begins to undergo a series of divisions or cleavages making 2, 4, 8, . . . 2^n cells.

The shape of the mass of daughter cells that emerges is usually that of a ball (although in birds it can be a sheet). Now I have to stop and consider some of the things cells do before I go ahead with this description:

1. Cells *divide*, passing on the same amount and kind of DNA to their daughter cells.
2. Cells *migrate*, separating from their connections in sheets called epithelia to form a loose, moving collection called a mesenchyme. (The sheets themselves can also move by curling up into tubes without releasing the contacts between their cells.)
3. Cells *die* in particular locations.
4. Cells *adhere* to each other, as we have already seen, or they lose their adhesion and migrate to another place. This migration occurs on the surfaces of other cells to form layers, or on matrix molecules released by the cells. The cells then readhere, forming new combinations.
5. Cells *differentiate*; they express different combinations of the genes present in their nuclei. They can do this at any time and place but only if they receive the right cues. Only certain places in the developing embryo have the right cues. This process of differential gene expression is called differentiation. It is what makes liver cells different from skin cells, and skin cells different from brain cells, and so on. Differentiation means specific patterns of protein production; some genes specifying particular proteins are turned on and some are turned off. Each cell of a given type has many proteins, only some of which are shared with cells of a different type.

Now we may resume our description of how an embryo is made. Let us consider the chick embryo (figure 6–3). Continued cell division eventually leads to a plate of cells called the blastoderm, which contains more than 100,000 cells. At this point, cells on either side of the midline in the posterior portion of the blastoderm detach and migrate through the middle portion, called the primitive streak. The result is that these cells end up beneath the blastoderm, where they adhere to form a layer called the mesoderm. Three separate layers—the ectoderm, mesoderm, and endo-derm—eventually form through this process, called gastrulation.

An amazing event that combines cell position and cell signaling occurs at this stage. This is called *embryonic induction*, and it is the result of signals passing from a set of cells in one layer to a set of cells in another. Cells in the mesoderm send signals to those in the ectoderm, resulting in a place-

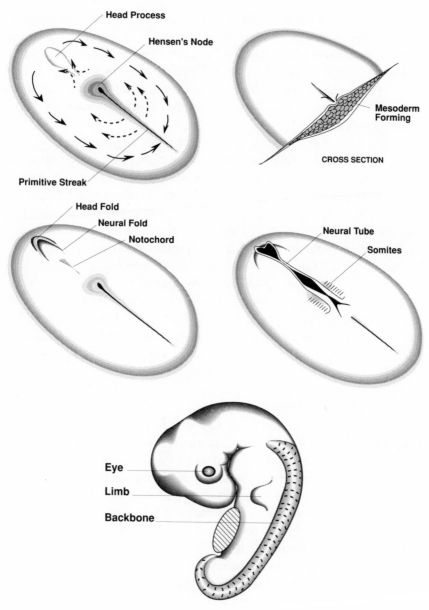

FIGURE 6–3

The early development of the chick embryo from a plate of cells that forms several layers in a process called gastrulation. Cells move through the primitive streak (top left) to form layers (top right). The central axis of the plate then folds into the neural tube, which will later give rise to the nervous system. Soon afterward, cells from the lower layers segregate (center left) and form segmented structures called somites (center right). The nervous system develops further as cells in the neural tube develop processes (shown in figure 3–3). The result is an embryo that begins to look like an individual animal (bottom).

dependent (topobiological) differentiation of a central section of cells to form the neural plate. Cells outside the boundary of the neural plate will form the skin; those of the plate itself will form the neural tube (figure 6–3) and subsequently the nervous system. They do so by rolling up as a sheet into the tube. This not only defines the axis of the embryo but sets the head end of the animal. It also sets the position-dependent cues for future induction events.

Notice how these primary cellular processes of division, migration, death, adhesion, and induction vary as a function of time and place. The critical step for consolidating these events is *coordinating* them to present new inductive signals leading to still further alterations in a particular place, one made as a result of past alterations. As a result of secondary inductions, for example, nerve cells in the neural tube send out fibers to form networks, each specific for a region of the brain or spinal cord (see figure 3–3). Other cells will form the eyes, gut, or kidneys. All of this occurs in such a fashion as to yield the shape characteristic of the species—in this case, a chick.

This matter of shape is critical. It means that *combinations* of genes act to give a heritable shape characteristic of that species. It also means that the mechanical events leading to the rearrangement and specialization of cells must be coordinated with the sequential expression of the genes. *This is the key requirement of topobiology.* It explains why genes specifying the shapes of proteins are not enough; individual *cells*, moving and dying in unpredictable ways, are the real driving forces. Making proteins or cell surfaces that latch on to each other, each specific for a given cell like a Lego toy, does not account for how genes specify shape. While the cells of an embryo of a species resemble each other *on the average*, the movement and death of a *particular* cell at any *particular* place is a statistical matter and that cell's actual position cannot be prespecified by the code in a gene.

What then *does* account for how shape is achieved in this marvelous sequence of cellular dances and signals? A clue lies with the molecules called morphoregulatory molecules that regulate adhesion and movement (figure 6–4). These proteins are specified by certain sets of genes at particular places in the embryo. Their main function is to cause cells to adhere or to link cells in sheets called epithelia. They fall into three families: cell adhesion molecules (CAMs) that link cells together directly, substrate adhesion molecules (SAMs) that link cells indirectly but provide a matrix or a basis on which they can move, and cell junctional molecules (CJMs) that link cells bound by CAMs into epithelial sheets.

The key point is this: The activation of genes for subsets of morphoregulatory molecules *modifies* the mechanics of cells and epithelia. This process is determined by the chemistry of each cell acting on the internal

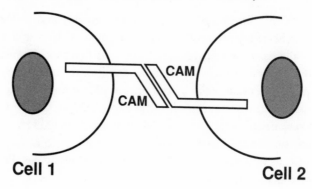

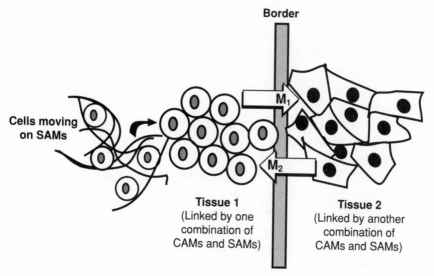

FIGURE 6–4

Cell adhesion. Top: Cells bind to each other by means of special proteisent in their outer membranes called cell adhesion molecules (CAMs). Bottom: Other molecules called substrate adhesion molecules (SAMs) form extracellular matrices on which cells move and rest. The CAMs and SAMs regulate how cells assemble and disassemble and permit or forbid movement. The processes that underlie the formation of shape, as shown in the previous figure, are under the control of special genes that are expressed in places at which molecular signals (morphogens, M_1 and M_2) are exchanged between adjoining collections of cells. The topobiological process is intricate, but the basic idea is simple: Cells move or stick at a particular place; after a set of genes is turned on or off, the cells are either released or kept and then produce new signals for new combinations. This results in further changes in CAM and SAM expression and the alteration of cell and tissue types by morphogens. These processes result in the shapes and the tissue types making up the embryo.

structures that affect cell shape and movement, and is therefore called mechanochemistry. A specific combination of CAMs and SAMs, for instance, allows some cells to move, controls mechanochemical events that fold the sheets formed when cells are linked, and even prevents the movement of certain cells in certain places. Because of these alternating permissive and restraining roles, the expression of CAMs that are linked to the surfaces of cells, and of SAMs that are deposited by cells to interact with the surfaces of other cells at particular places, can alter the combinations of cells at a given place, leading to different shapes. Particular genes control the formation of particular CAMs and SAMs, so this role can be inherited for a given species, thus determining its shape-forming mechanism.

Even this is not enough, however. Along with CAMs, SAMs, and CJMs expressed at specific places in specific amounts and kinds, other events must also occur. After a place is constructed by inductive signals, new signal combinations must tell cells in that place which new CAMs to turn on and which old ones to turn off. These signals consist of small molecules or growth factors that interact either directly or indirectly with the appropriate genes to control this expression (figure 6–4).

A similar thing must happen to cause cells to differentiate or not to differentiate in the right place. Molecular biologists have identified a special kind of gene known as a homeotic gene. The homeotic gene specifies proteins that bind to portions of other genes and subsequently regulate the production of the proteins specified by those genes. In this manner, homeotic genes control the differentiation events that make a body region such as a wing, or an eye, or a part of the vertebral column. In a fruit fly, for example, a mutation in a homeotic gene can cause a leg to grow where an antenna should be (figure 6–5). Homeotic genes are expressed in gradients across the animal, usually front to back, and in particular regions.

To summarize: Cells express genes in time and space to govern morphoregulatory molecules, which in turn control cell movements and cell-to-cell adhesion. These actions place groups of cells in proximity, allowing them to exchange further inductive signals. These alter the expression of homeotic genes, which then alter the expression of other genes. The key players in this topobiological cascade are the cells, which move, die, divide, release inductive signals or morphogens, link to form new sheets, and repeat variants of the process. Genes control the whole business indirectly by governing which morphoregulatory or homeotic product will be expressed. But the actual microscopic fate of a cell is determined by epigenetic events that depend on developmental histories unique to each individual cell in the embryo.

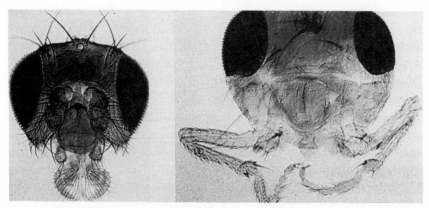

FIGURE 6–5

An example of aberrant topobiology. Homeotic genes, which regulate other genes, are expressed in particular places during the development of the embryo. If a homeotic gene undergoes a mutation, a body part such as a leg can replace an antenna in the fruit fly Drosophila. *The picture on the left shows a normal fly head; the one on the right shows a mutant fly with legs where the antennae should be.*

The end result is morphology. And because each species has a particular combination of genes, the frequency of which is established on the average by natural selection during evolution, there is a shape or tissue structure more or less characteristic of that species.

This account provides a provisional answer to the question of how a one-dimensional genetic code can specify a three-dimensional animal. It also suggests how evolution leads to relatively large and rapid changes in morphology. Suppose, for example, that, during evolution, a mutation affected the timing or binding of a morphoregulatory molecule, delaying its expression in sufficient amounts until cells in the embryonic region had divided more than usual. A larger structure might result, with a different shape. If animals with the new size and shape showed increased fitness in a given environment, natural selection would lead to differential reproduction of these animals. This would result in an increase in the frequency of the mutant gene in that population, and more animals would be born with the variant size and shape.

Why have I gone into this degree of detail? The reason is twofold: The nervous system and the brain are formed by such processes as are described here, and the signaling that occurs in the nervous system is topobiological (see chapter 3). The maps of the nervous system that result from nerve cells sending their processes to other regions of cells during development are among the most remarkable of topobiological structures. Their formation has been shown to depend on morphoregulatory molecules. Moreover, maps often depend on the selective death of the cells that compete to make them. They also require special signaling processes that locate the branches of active neighboring nerve cells in such a way that they end up as neighbors again in the distant map that is their target. Evidence suggests that if the chemical or electrical activity of these cells is blocked, their branches will not form orderly maps in some distant place (see figure 3–4).

Imagine now this epigenetic drama in which sheets of nerve cells in the developing brain form a neighborhood. Neighbors in that neighborhood exchange signals as they are linked by CAMs and CJMs. They send processes out in a profuse fashion, sometimes bunched together in bundles called fascicles. When they reach other neighborhoods and sheets they stimulate target cells. These in turn release diffusible substances or signals which, if the ingrowing processes have correlated signals, allow them to branch and make attachments. Those that do not either pass on or retract. Indeed, if they do not meet their targets, their parent cells may die. Finally, as growth and selection operate, a mapped neural structure with a function may form. The number of cells being made, dying, and becoming incorporated is huge. The entire situation is a dynamic one, depending on signals, genes, proteins, cell movement, division, and death, all interacting at many levels.

Notice the main features of this drama. It is topobiological, or place-dependent. Events occurring in one place require that previous events have occurred at other places. But it is also inherently dynamic, plastic, or variable at the level of its fundamental units, the cells. Even in genetically identical twins, the exact same pattern of nerve cells is not found at the same place and time. Yet the collective picture is species-specific because the *overall* constraints acting on the genes are characteristic of that species.

The events I have described are selectional ones. Certain patterns of cells are selected from a variant mass of cells in a topobiological fashion. This is dramatically the case in the nervous system. Selection not only guarantees a common pattern in a species but also results in *individual* diversity at the level of the finest neural networks. I have already mentioned that the diversity or variability of the connections at a given place in the nervous system argues against the idea that the brain functions like a computer. Diversity must inevitably result from the *dynamic* nature of topobiological events. The existence of diversity at the level of the individual animal is of great importance. Indeed, it is likely to be one of the most important features of the morphology that gives rise to mind.

But we are not there yet! First, we have to ask how biological systems carry out recognition events—how, without the transfer of preexisting, specifically coded messages, a biological system nonetheless specifically distinguishes one thing from another.

CHAPTER 7

The Problems Reconsidered

Man considering himself is the great prodigy of nature. For he cannot conceive what his body is, even less what his spirit is, and least of all how body can be united with spirit. That is the peak of his difficulty and yet it is his very being.

—Blaise Pascal

Putting the mind back into nature has precipitated a series of scientific crises, for the data on the brain, mind, and behavior do not correspond to the pictures we have been using to explain them. Many people think this is an audacious conclusion—unwarranted, premature (more facts will clear things up!), or even downright unhealthy. I think, on the contrary, that the best time to be working in a science is when it is in a crisis state. It is then that one is prompted to think of a new way of looking at the data, or of a new theory, or of a new technique to resolve an apparent paradox. One of the most striking crises of modern science occurred, for example, when it was understood that the application of the classical laws of physics to a heated metal block with a cavity (a "black body radiator") led to an impossible situation at short wave lengths and high energies; in this so-called ultraviolet catastrophe, energy becomes infinite. The solution compelled by this situation was given by Max Planck, who suggested that energy was not radiated by a hot body continuously but rather in packets or quanta.

The crises in considering the matter of the mind are in no way as clear-cut, however. This should come as no surprise, given how subtle and multilayered the business of brain development, brain action, and mental activity is. It begins with molecules and goes on to genes. It involves vast numbers of cells with electrical activity and chemical diversity, an enormously intricate anatomy with blobs and sheets linked in rich ways, and maps that receive signals from sensory input and send signals to motor output. These structures undergo continuous electrical and chemical change, driving and being driven by animal movement. This movement is itself conditioned by animal shape and pattern, leading to behavior. Some of this behavior involves communication with an animal's memory, which is in turn affected by its own products.

All of this comes about as a result of evolution—that is, as a result of natural selection operating over hundreds of millions of years. It is no wonder that the crises of brain science and psychology are not as neat or as evident as those of physics. The sheer complexities are much greater than those of physics. Yet as I have shown (see chapter 3), the crises emerge in stark clarity provided that one is willing to reach across the different levels in an effort to relate structure to function.

What can we do to reconcile the interactions of the various levels and to resolve the crises of structure and function that they jointly pose? The answer lies in seeing what the critical problems are, avoiding category errors, and constructing a theory. Of course, this theory must be scientific—that is, it must be testable or falsifiable by experimental means. But it need not *always* lead to predictions at all of its levels, nor need every part of it be immediately or obviously falsifiable. (Had such strict criteria of falsifiability been applied to Darwin's evolutionary theory at its inception, it would have been prematurely abandoned.)

The next part of this book will summarize such a theory, already described at much greater length in my trilogy of books on morphology and mind (see the Selected Readings at the back of this book). I will give the gist of what is contained in that set of volumes. I plan to simplify the task by first describing some known biological systems that have properties analogous to those of the brain. But I want to warn the reader that these analogies are heuristic; they are intended to help with the comprehension of certain mechanisms, not to be explicit examples of cognitive functions.

Before turning to these analogies, it may be useful to reconsider the problems we started with and to summarize the argument thus far. As long as science and scientific observers dealt with physical objects and natural forces independent of the minds of the observers, a grand set of theories within a group of compatible sciences could afford to ignore the psychological intricacies of scientific observers. While their sensations and perceptions went into the performance of their experiments and into intersubjective exchanges with their colleagues, these sensations and perceptions were strictly excluded from their theoretical and formal explanations. Aside from a few difficulties at the boundaries of the very small (in quantum measurement) or of the very fast or large (in relativity theory), the scientific observers' participation *appeared* to be from a God's-eye view. An "objectivist" picture of nature developed that distinguished things from each other by "classical categories": categories defined by singly necessary and jointly sufficient conditions. These were then mapped onto the physical world in an unambiguous fashion by incorporating experimental data into far-reaching physical theories.

In many domains, this approach worked enviably well (and still does), but when the mind was put back into nature by nineteenth-century studies of physiology and psychology, a series of difficulties began to emerge. One of the first of these difficulties was that the observer could no longer neglect mental events and mental experience. He could no longer ignore consciousness itself or the fact that conscious experience was intentional— always in reference to an object. The mechanisms of this consciousness were not directly transparent, nor could consciousness be studied directly as an external object—at best, it could be introspected or indirectly inferred from the behavior of others.

One reaction to this state of affairs was to declare the subject off limits and insist that science should concern itself only with behavior that was observable in ways defined by the forms of successful scientific inquiry concerned with nonintentional objects. In an attempt to salvage the "scientific" posture without denying intentionality, and in contrast to this behaviorism, a different position was later taken by cognitive science. The cognitive position was to adopt notions derived from logical and formal analysis, putting an emphasis on syntax. In this view, the mind, like a computer, is organized by rules and operates by mental representations. Meanings or semantics are supposed to arise by mapping these rules onto classically categorizable events and objects. Unlike behaviorism, this view allowed one to look into the mind but then described it as if it were a formal system. This description floated more or less free of the detailed structure of the brain. The semantic mapping of that description onto the world is objectivist; things and events are unequivocally described as classical categories.

As I discuss in the Postscript, however, proposals that the brain and the mind function like digital computers do not stand up to scrutiny. The idea of mental representations posited without reference to brain mechanisms and structures does not fare much better. An examination of how animals and people categorize the world, and how babies mentally develop, undercuts the idea that language can be adequately explained by syntactical analyses carried out in the absence of an adequate explanation of meaning. The objectivist view of the world is at best incomplete and at worst downright wrong. The brain is not a computer and the world is not a piece of computer tape.

As a young scientist, I believed that physics *would* explain and exhaust everything, at least someday. I didn't know it, but I was an objectivist. Now, while my regard for physics remains as high, I see that some supplementation is required to get intentionality into the picture. My present view of the structure of things is exemplified by the story of the gentleman who had the paranoid delusion that his girlfriend was seeing another suitor. One hot summer night he came home early to their apartment and in a fury of jealousy searched everywhere for the hypothetical suitor, but could not find him. Still in his rage, he found himself at the back window of the apartment. He looked out at the fire escape below, and saw a man loosening his collar and wiping his brow. Flying into a greater rage, he lifted a very large refrigerator, fit it through the window, aimed it carefully, and dropped it on the man's head—whereupon he fell dead of exertion.

The scene switches to Heaven. Three people are being admitted; Saint Peter tells them that they have fulfilled all the requirements but the last, which is to describe the nature of their deaths. The first man said, "Well, I thought there was some hanky-panky going on, so I came home early. I looked all over the place and finally found this fellow, and I must have had an adrenaline fit. I lifted a refrigerator I could not ordinarily lift, dropped it on his head, and must have had a heart attack." The second man said, "I don't know. It was a hot summer night. I stepped out onto the fire escape, loosened my collar and wiped my brow, and a refrigerator fell on my head." The third man said, "I don't know. I was just sitting in this refrigerator, minding my own business."

The physics of falling bodies, certainly, and also of intentionality, underlaid by some rather critical morphology in the head, none of which can be disturbed by unverified fantasies *or* by heavy objects without serious consequences.

The notion that we can think about how mental matters occur in the absence of reference to the structure, function, development, and evolution of the brain is intellectually hazardous. The likelihood of guessing how the brain works without looking at its structure seems slim. Certainly, if one agrees with the ethologists that mental states are a product of evolution, we must at least study how the brain evolved. Our obligation is to complete Darwin's program.

When we make even our first halting efforts to do so, we come upon a series of intriguing and baffling findings. We see that the development of brains is enormously dynamic and statistical. Developmental analysis suggests that the way genes regulate the intricate anatomy of the brain is through epigenetic interactions—particular developmental events must occur before others can occur. Certain adhesion molecules regulate collectives of cells and their migration, but do not do so cell by cell in a prescribed or prearranged pattern. And to some extent, cell migration and cell death are stochastic—they have unpredictable consequences at the level of individual cells. These statistical processes oblige individual brains, unlike computers, to be individual. The somatic diversity necessarily generated by these means is so large that it cannot be dismissed as "noise," as one would dismiss the noise in an electronic circuit at normal operating temperatures. (The hiss from your hi-fi amplifier is an example.)

Indeed, the circuits of the brain look like no others we have seen before. The neurons have treelike arbors that overlap and ramify in myriad ways. Their signaling is not like that in a computer or a telephone exchange; it is more like the vast aggregate of interactive events in a jungle. And yet despite this, brains give rise to maps and circuits that automatically adapt their boundaries to changing signals. Brains contain multiple maps interacting without any supervisors, yet bring unity and cohesiveness to perceptual scenes. And they let their possessors (pigeons, for example) categorize as similar a large if not endless set of diverse objects, such as pictures of different fish, after seeing only a few such pictures.

If you consider these extraordinary brain properties in conjunction with the dilemmas created by the machine or the computer view of the mind, it is fair to say that we have a scientific crisis. The question then arises as to how to resolve it. For a possible way out, let us look to biology itself rather than to physics, mathematics, or computer science.

PART III

PROPOSALS

We are now in a position to use what we know about biology, psychology, and philosophy to postulate a theory of consciousness that will be an essential part of a theory of how the brain works.

Most scientists consider such efforts premature, if not downright crazy. But the history of science suggests that we progress not by simply collecting facts but by synthesizing ideas and then testing them. It also teaches us that nothing is so effective in promoting new thoughts and experiments as a theory that one can amend or even knock down.

The theory must be a scientific one: Its parts must be testable, and it must help to organize most, if not all, of the known facts about brains and minds. To accomplish this for the matter of the mind means sifting through several layers of organization in the nervous system.

It also means rethinking what we mean by "memory," "concepts," "meaning," "conscious as an animal," and "conscious as a human being." At the very least, in attempting to do so, the reader will learn about some fascinating biological phenomena and findings. At the most, he or she may glimpse a view of the material bases of mind. The reader is urged to have patience—we are at the frontier, a place where boundaries shift, where although amenities may be lacking, the sense of excitement is heightened.

C H A P T E R 8

The Sciences of Recognition

Selection is better than instruction.

—Anonymous

At this point, we are prepared to approach the matter of the mind from a biological point of view. This is less a matter of biochemistry, cell biology, or neurophysiology than it is a conceptual mode—a way of looking at mental matters from a vantage point based on biological concepts.

It is not commonly understood that there are characteristically biological modes of thought that are not present or even required in other sciences. One of the most fundamental of these is population thinking, developed largely by Darwin. Population thinking considers variation not to be an error but, as the great evolutionist Ernst Mayr put it, to be real. Individual variance in a population is the source of diversity on which natural selection acts to produce different kinds of organisms. This contrasts starkly with Platonic essentialism, which requires a typology created from the top down; instead, population thinking states that evolution produces classes of living forms from the bottom up by gradual selective processes over eons of time (see figure 5–2). No such idea exists in physics—even the "evolution" of stars does not require such a notion for its explanation.

Connecting population thinking to ideas of the mind, I have to go a bit deeper into some of its consequences and into those of the evolutionary process itself. To do so, I will call on some terms that are probably new to the reader, at least in their specialized usage. Among them are "instruction," "selection," "recognition," and "memory" or "heritability." Notice that, except for the last, all of these words are commonly used, but for our purposes, I will use them somewhat differently.

Let me begin with the specialized notion of recognition. I will make an abstract statement and then translate it into an evolutionary example. By "recognition," I mean the continual adaptive matching or fitting of elements in one physical domain to novelty occurring in elements of another, more or less independent physical domain, a matching that occurs without prior instruction. This is quite a mouthful; let me see if I can cut it into digestible morsels by using evolution as an example.

In evolution, organisms (elements of domain 1) are more or less adapted to events in the environment (elements of domain 2). This adaptation occurs even when environmental changes cannot be predicted (that is, even when the changes represent novelties). The process of adaptation occurs by selection on those organismal variants that are on the average fittest, and what makes them fittest does not require *prior* explicit information ("instruction") about the nature of the novelties in the environment. The selective environmental changes are, in general, independent of variation in the population of organisms, although selection resulting from such changes may add to that variation. In sum, there is no explicit information transfer between the environment and organisms that causes the population to change and increase its fitness. Evolution works by selection, not by instruction. There is no final cause, no teleology, no purpose guiding the overall process, the responses of which occur *ex post facto* in each case.

This is an astonishing idea. It reminds me of the lady in the E. M. Forster novel who said, "How do I know what I think until I see what I say?" Even more astonishing is the fact that evolution, acting by selection on populations of individuals over long periods of time, gives rise to selective systems *within* individuals. Such selective systems acting in one lifetime in one body are called somatic selective systems. Thus, an evolutionary selective system selects for a somatic selective system!

Now I will describe a specific example, the immune system, which is seen only in backboned animals. Grasping the fundamentals of immunity will be very useful in understanding selection in the nervous system. This description will, I hope, justify my rather specialized and abstract use of the word "recognition," for the immune system is the best understood example of a somatic recognition system based on selectionist principles. Indeed, before I am through, I want to make the case that there are sciences of recognition, sciences that study recognition systems. The evidence is abundant that evolution and adaptive immunity are two such systems acting over different populations and over different time scales. This is by way of leading up to the suggestion that the brain also acts as a somatic selective system and thus that neurobiology is also a science of recognition.

But I am getting ahead of myself. I want to describe briefly the immune system, the science of which I worked in for some fifteen years, for it is both intriguing and illuminating. The immune system is a somatic selective system consisting of molecules, cells, and specialized organs. As a system, it is capable of telling the difference between self and nonself at the molecular level. For example, it is responsible for distinguishing between and responding to the chemical characteristics of viral and bacterial invaders (nonself), invaders that would otherwise overwhelm the collections of cellular systems in an individual organism (self). This response involves molecular recognition with an exquisite degree of specificity. An appropriately stimulated immune system can tell the difference between two large foreign protein molecules composed of thousands of carbon atoms that differ by only a few degrees in the tilt of a single carbon chain. It can tell these molecules apart from all other molecules and retain the ability to do so once it has initially developed that ability. It has a "memory."

If I inject a protein into an individual's body that does not resemble its own proteins, specialized cells called lymphocytes respond by producing molecules called antibodies (figure 8–1). These molecules bind by fitting to specific and characteristic portions of the foreign molecule, or antigen, as it is called. A second and later encounter allows these antibodies to bind even more effectively to just those antigens. Perhaps more astonishing is the fact that a specific recognition event occurs even for new molecules synthesized by organic chemists, molecules that never existed before either in the responding species or in the history of the earth for that matter.

How can an individual's body positively distinguish novel molecules in such a specific fashion? The theory prevailing before the present one was called the theory of instruction. Its basic assumption was that, in the immune system, a foreign molecule transferred information about its shape and structure to the combining site of the antibody molecule. It then removed itself (the way a cookie cutter would be removed from dough) leaving a crevice of complementary shape that could then bind to all foreign molecules with regions having the shape with which the impression was originally made. It is obvious why this is an instructive process: Information about a three-dimensional structure is posited to be *necessary* to instruct the immune system how to form an antibody protein whose polypeptide chain folds around that structure (see figure 6–1, bottom) to give the appropriate complementary shape.

This elegant and simple theory turned out to be false. The theory that

ANTIBODY MOLECULE

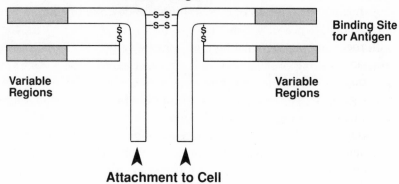

Constant Regions

Binding Site
for Antigen

Variable
Regions

Variable
Regions

Attachment to Cell

CLONAL SELECTION

Lymphocyte Repertoire

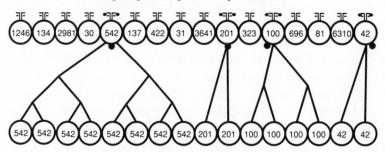

Clonal Cell Division

FIGURE 8–1

The immune system works as a selective recognition system. Your immune system distinguishes foreign molecules (nonself) from the molecules of your body (self) by virtue of their different shapes. It does so by making proteins called antibodies. Each immune cell makes an antibody with a different variable region (top figure); each variable region has a binding site with a different shape. When a foreign molecule or antigen (bottom figure, black dots) enters the body, it is bound by just those antibodies on the cells of the immune system that happen to fit parts of its shape (cells 542, 201, 100, and 42). This set of cells then divides and makes a "clone"—more cells of the same kind with antibodies of the same kind. The next time the antigen is presented, many more copies of these same antibodies are there to help destroy it. Cells numbered 542, 201, 100, and 42 are now more prevalent, for example, and will recognize the foreign molecules more rapidly the next time they intrude. The foreign intruders could be molecules on a virus or a bacterium.

replaced it is more complicated and even belies common sense, but it appears to explain a wide variety of facts; indeed, few, if any, present-day immunologists would dispute its essential correctness. This theory is known as the theory of clonal selection and was first proposed by the late Sir Frank MacFarlane Burnet.

Burnet maintained that, prior to a confrontation with any foreign molecule, an individual's body has the ability to make a huge repertoire of antibody molecules, each with a different shape at its binding site. When a foreign molecule (say on a virus or bacterium) is introduced into the body, it encounters a *population* of cells, each with a *different* antibody on its surface. It binds to those cells in the repertoire having antibodies whose combining sites happen to be more or less complementary to it. When a portion of an antigen binds to an antibody with a sufficiently close fit, it stimulates the cell (called a lymphocyte) bearing that antibody to divide repeatedly (figure 8–1). This results in many more "progeny" cells having antibodies of the same shape and binding specificity.

A group of daughter cells is called a clone (the asexual progeny of a single cell) and the whole process is one of differential reproduction by clonal selection. In other words, as a result of the selection of cells with appropriately specific antibodies from a large repertoire of diverse cells, the kinds of antibodies specific for a foreign shape are increased in number because the selective binding event *ex post facto* caused those cells to multiply. The composition of the lymphocyte population has been changed by selection.

An analysis of the complete structure of an antibody was carried out in my laboratory several decades ago. It showed that the polypeptide chains of an antibody (figure 8–1, top) consist of constant regions (similar or identical from molecule to molecule) and variable regions (different for each kind of molecule and comprising the binding site.) We now know that this diversity is generated somatically (that is, within an individual's lifetime) in the lymphocytes of each individual's body. The process involves a kind of jumbling within each lymphocyte of the genetic code specifying the variable antibody regions that *may* someday happen to bind an antigen.

I hope I have said enough to show you that the immune system

This is a selective system because vast numbers of different antibody-binding shapes are present (each one on a different cell) before the antigens enter. These antigens select only a few of the shapes, and antibody production is vastly amplified by clonal division of the cells (2,4,8,16. . . .) to enormous numbers. Thus, the population is changed as a result of experience.

corresponds to my definition of a recognition system. It exists in one physical domain (an individual's body) and responds to novelty arising independently in another domain (a foreign molecule among the millions upon millions of possible chemically different molecules) by a specific binding event and an adaptive cellular response. It does this *without* requiring that information about the shape that needs to be recognized be transferred to the recognizing system *at the time when it makes the recognizer molecules or antibodies.* Instead, the recognizing system *first* generates a diverse population of antibody molecules and then selects *ex post facto* those that fit or match. It does this continually and, for the most part, adaptively.

The immune selective system has some intriguing properties. First, there is more than one way to recognize successfully any particular shape. Second, no two individuals do it exactly the same way; that is, no two individuals have identical antibodies. Third, the system has a kind of cellular memory. After the presentation of an antigen to a set of lymphocytes that bind it, some will divide only a few times, while the rest go on irreversibly to produce antibody specific for that antigen and die. Because some of the cells have divided but not all the way to the antibody-making end, they constitute a larger group of cells in the total population of cells than were originally present. This larger group can respond at a later time in an accelerated fashion to the same antigen. As I mentioned before, the system therefore exhibits a form of memory at the cellular level.

Here is a molecular recognition system that is noncognitive and highly specific, the explanation of which is a marvelous example of population thinking—the essence of Darwinism. Like evolution, it has a generator of diversity (the "jumbler" of DNA in each lymphocyte), a means of perpetuating changes by a kind of heredity (clonal division), and a means of differentially amplifying selection events (differential clonal reproduction). Unlike evolution, it occurs in *cells* over *short periods of time* and does not produce many levels of form—just different antibody molecules. It is a somatic selective system.

Notice that in evolution itself, diversity is generated within a population of *animals* by mutations in DNA. These are transmitted hereditarily through germ cells (sperm and egg). Selection then occurs on individuals continually over *evolutionary* time to give rise to different species, depending on many variables in the environment. Both systems, evolution and immunity, deal with novelty by similar selective principles but by very different mechanisms. It is conceptually important to distinguish a selective principle from the mechanisms used to express it in any particular physical system.

What these two sciences of recognition, evolution and immunology, have in common is not found in nonbiological systems such as "evolving" stars. Such physical systems can be explained in terms of energy transfer, dynamics, causes, and even "information transfer." But they do not exhibit repertoires of variants ready for interaction by selection to give a population response according to a hereditary principle. The application of a selective principle in a recognition system, by the way, does not *necessarily* mean that genes must be involved—it simply means that any state resulting after selection is highly correlated in structure with the one that gave rise to it and that the correlation continues to be propagated. Nor is it the case that selection cannot itself introduce variation. But a constancy or "memory" of selective events *is* necessary. If changes occurred so fast that what was selected could not emerge in the population or was destroyed, a recognition system would not survive. Physics proper does not deal with recognition systems, which are by their nature biological and historical systems. But all the laws of physics nevertheless apply to recognition systems.

Leo Szilard, a great physicist whose experiments with Enrico Fermi led to controlled nuclear fission, was fascinated with both immunology and the brain. He used to visit my laboratory often to see what was new with antibodies. Usually, he would start by saying, "All right, what's the problem? I have fifteen minutes." Once Szilard attended a meeting at which an unfortunate researcher proposed a theory of memory. Thought, he explained, caused new proteins to be made by our brains. After a time, these being new, they would stimulate antibodylike proteins that represented memories. Leo rose and said, with a merry and merciless smile, "Maybe that's how *your* brain works."

Do brains constitute selective recognition systems? Will describing the fundamental operations of brains in such terms be revealing and useful? As you have undoubtedly surmised, I think it will not only be useful, but it will also remove much of the paradox and sense of crisis that one confronts when reviewing the data on brain structure and function. Indeed, I believe that neurobiology is a science of recognition. But even though after our antibody work my colleagues and I were excited to discover that neural cell adhesion molecules or "brain glue" are the evolutionary precursors of the whole immune system, I am hardly suggesting the kind of proposal made by Szilard's unhappy correspondent. The resemblance between the immune and nervous systems is only in principle, not in detail.

I have defended the notion that brains are selective recognition systems because thinking about brain function in selectional terms relieves us of the horror of the homunculus (figure 8–2). Because diversity exists beforehand

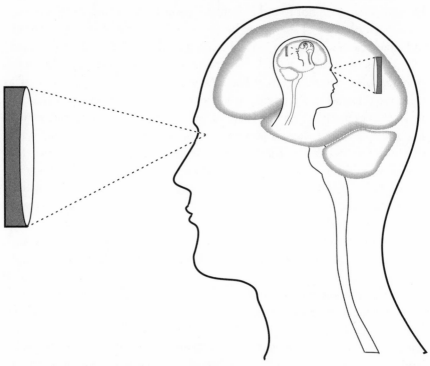

FIGURE 8–2

The endless regression of homunculi. The idea of instruction or information processing requires someone, or something, to read it. But a similar entity is then required to read the resulting messages, and so on, endlessly.

in a selective system, and because specificity arises as a result of selection *ex post facto*, we are no longer faced with an endless regress of information processors in the head. To defend this statement adequately, I will describe a theory of brain function that follows selectional principles. The challenge is to show how evolution and development give rise to a somatic selectional system in the brain. Having accomplished that, I will attempt to show that the selectional mechanisms proposed can account for psychological functions—perception, memory, even consciousness. Let us turn to the various parts of this task.

CHAPTER 9

Neural Darwinism

If a term has to be used for the whole set of ideas I would suggest Neural Edelmanism.

—Francis H.C. Crick

I f we consider recognition to be a kind of adaptive matching, then it is obvious why it applies to evolution and immunity. In both instances, population thinking provides a means for understanding. What is the justification for applying population thinking to the workings of the brain, for neural Darwinism? This is not the place to go into all the intricacies of such a position, but I believe it will clarify much of what I have to say in the rest of this book if I give some of the reasons for considering brain science a science of recognition.

The first reason is almost too obvious: Brain science and the study of behavior are concerned with the adaptive matching of animals to their environments. In considering brain science as a science of recognition I am implying that recognition is not an instructive process. No direct information transfer occurs, just as none occurs in evolutionary or immune processes. Instead, recognition is selective.

Some justifications for this position may be found in my previous criticisms (chapter 3) of various category mistakes in thinking about the brain; an extended argument supporting these criticisms may be found in the Postscript. I have already argued that the world is not a piece of tape and that the brain is not a computer. If we take such a position, we have to show how a behaving animal nevertheless adaptively matches its responses to unforeseen novelty occurring in such a world. There is an additional set of reasons for assuming that recognition cannot be instructive. We have already seen that the individuality and structural diversity of brains even within one species is confounding to models that consider the brain to be a computer. Evidence from developmental studies suggests that the extraordinary anatomical diversity at the finest ramifications of neural networks is an unavoidable consequence of the embryological process. That degree of individual diversity cannot be tolerated in a computer system following instructions. But it is just what is needed in a selective system.

A potent additional reason for adopting a selective rather than an instructive viewpoint has to do with the homunculus. You will remember that the homunculus is the little man that one must postulate "at the top of the mind," acting as an interpreter of signals and symbols in any instructive theory of mind. If information from the world is processed by rules in a computerlike brain, his existence seems to be obliged. But then another homunculus is required in *his* head and so on, in an infinite regress (see figure 8–2). Selective systems, in which matching occurs *ex post facto* on an *already existing* diverse repertoire, need no special creations, no homunculi, and no such regress.

If we assume that brain functions are built according to a selectional process, we must be able to reconcile the structural and functional variability of the brain with the need to explain how it carries out categorization. To do so, we need a theory with a number of essential characteristics. It must be in accord with the facts of evolution and development; account for the adaptive nature of responses to novelty; show how the brain's functions are scaled to those of the body as the body changes with growth and experience; account for the existence and functions of maps in the brain— why they fluctuate, how multiple maps lead to integrated responses, and how they lead to generalizations of perceptual responses, even in the absence of language. Eventually, such a theory would also need to account for the emergence of language itself. And finally, such a theory must account for how the various forms of perceptual and conceptual categorization, of memory, and of consciousness arose during evolution.

To be scientific, the theory must be based on the assumption that all cognition and all conscious experience rest solely on processes and orderings occurring in the physical world. The theory must therefore take care to explain how psychological processes are related to physiological ones.

The theory I have proposed to account for these matters is known as the theory of neuronal group selection (TNGS). Its basic tenets are described in this chapter and those of its features that provide a bridge between psychology and physiology are stressed. This will enable us to come to grips with the daunting problem of consciousness, and it is one of the main reasons for explaining the theory in any detail. In the course of doing so, I deal with perceptual categorization, concepts, memory, and learning.

I want to warn the reader that I have to explain a series of complex processes that must be grasped in order to understand brain function. The main ideas to grasp are called neuronal group selection, reentry, and global mapping. I provide examples for each. If they are mastered, they will serve you well, because we will use them over and over again in later chapters.

In the sense of requiring all these processes, the TNGS is a complex theory; its basic tenets, however, are only three in number. No additional tenets are required to explain even so remarkable a property as consciousness. What *is* required to explain such a property, however, is the evolution of new kinds of morphology in both the body and the brain. So I will take up some features of these morphologies as we go along.

The three tenets of the TNGS (figure 9–1) are concerned with how the anatomy of the brain is first set up during development, how patterns of responses are then selected from this anatomy during experience, and how reentry, a process of signaling between the resulting maps of the brain, gives rise to behaviorally important functions.

According to the first tenet, developmental selection, the dynamic primary processes of development discussed in chapters 3 and 6 lead to the formation of the neuroanatomy characteristic of a given species. This anatomy obligatorily possesses enormous variation at its finest levels and ramifications. This is because of the dynamic regulation of CAMs and SAMs, the stochastic fluctuation of cell movement, cell process extension, and cell death during development, and the activity-dependent matching of connections that is superimposed on neural branches (or neurites) as they explore a developing brain region. This entire process is a selectional one, involving populations of neurons engaged in topobiological *competition*. A population of variant groups of neurons in a given brain region, comprising neural networks arising by processes of somatic selection, is known as a *primary repertoire*. The genetic code does not provide a specific wiring diagram for this repertoire. Rather, it imposes a set of *constraints* on the selectional process. Even with such constraints, genetically identical individuals are unlikely to have identical wiring, for selection is epigenetic.

The second tenet of the TNGS provides another mechanism of selection that, in general, does not involve an alteration of the anatomical pattern. It assumes that, during behavior, synaptic connections in the anatomy are

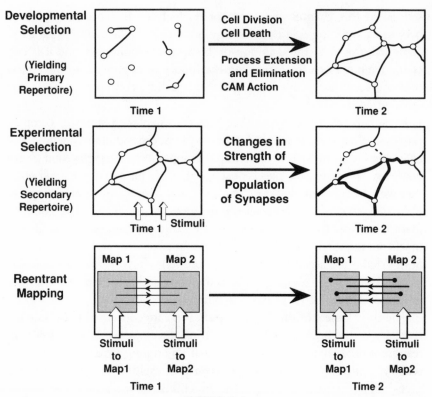

FIGURE 9–1

A selectional theory of brain function. Called the theory of neuronal group selection, it has three tenets. Top: Developmental selection. *This occurs as a result of the molecular effects of CAM and SAM regulation, growth factor signaling, and selective cell death to yield varied anatomical networks in each individual, networks that make up the primary repertoire.* Center: Experiential selection. *Selective strengthening or weakening of populations of synapses as a result of behavior leads to the formation of various circuits, a secondary repertoire of neuronal groups. The consequences of synaptic strengthening are indicated by bold paths; of weakening, by dashed paths.* Bottom: Reentry. *The linking of maps occurs in time through parallel selection and the correlation of the maps' neuronal groups, which independently and disjunctively receive inputs. This process provides a basis for perceptual categorization. Dots at the ends of the active reciprocal connections indicate parallel and more or less simultaneous strengthening of the synapses in reentrant paths. (Readers may wish to refresh their knowledge of synapses by referring to figure 3–2.) Strengthening (or weakening) can occur in both intrinsic and extrinsic reentrant connections.*

selectively strengthened or weakened by specific biochemical processes. This mechanism, which underlies memory and a number of other functions, effectively "carves out" a variety of functioning *circuits* (with strengthened synapses) from the anatomical network by selection. Such a set of variant functional circuits is called a *secondary repertoire.*

To some extent, the mechanisms leading to the formation of primary and secondary repertoires are intermixed. This is so because at certain times and places the formation of the primary repertoire depends on changing synaptic strengths, as in the activity-dependent matching of connections (for an example, see figure 3–4). Even in a developed brain, "sprouting" can occur, in which new neural processes form additional synapses. And in some cases, such as the development of bird song and frog metamorphosis, the formation of new parts of the nervous system involving simultaneous primary and secondary repertoire formation occurs during behavior in the world.

The third tenet of the TNGS is concerned with how the selectional events described in the first two tenets act to connect psychology to physiology. It suggests how brain maps interact by a process called reentry. This is perhaps the most important of all the proposals of the theory, for it underlies how the brain areas that emerge in evolution coordinate with each other to yield new functions.

To carry out such functions, primary and secondary repertoires must form maps. These maps are connected by massively parallel and reciprocal connections. The visual system of the monkey, for example, has over thirty different maps, each with a certain degree of functional segregation (for orientation, color, movement, and so forth), and linked to the others by parallel and reciprocal connections (figure 9–2). Reentrant signaling occurs along these connections. This means that, as groups of neurons are selected in a map, other groups in reentrantly connected but different maps may also be selected at the same time. Correlation and coordination of such selection events are achieved by reentrant signaling and by the strengthening of interconnections between the maps within a segment of time.

A fundamental premise of the TNGS is that the selective coordination of the complex patterns of interconnection between neuronal groups by reentry is the basis of behavior. Indeed, reentry (combined with memory, which I discuss later) is the main basis for a bridge between physiology and psychology.

I have not yet mentioned what the unit of selection is for the brain. In evolution, the main unit of selection is the individual animal (the phenotype) and in immunity, it is the individual lymphocyte. According to the TNGS, however, the unit of selection is not the individual nerve cell, but

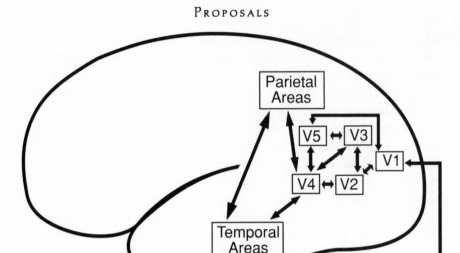

FIGURE 9–2

Multiple maps of visual areas of the brain are reentrantly connected to each other (see double arrows linking visual maps V1–V5, the temporal areas, and the parietal areas). Each map serves in a functionally segregated manner—for color, motion, orientation, and so forth. No "supervisor map" exists that summarizes "information" on these properties. But as a result of reentry (the double arrows), the maps act coherently to respond to combinations of properties. Even the region known as the lateral geniculate nucleus (LGN), an infracortical region that receives signals from the optic nerve of the eye, is reentrantly connected to the primary visual map, V1.

is rather a closely connected collection of cells called a neuronal group (figure 9–3). The reason for this has to do with limitations on the properties of neurons, with developmental constraints, and with neuroanatomical requirements on reentrant circuits. Individual neurons are either excitatory for other neurons or inhibitory for other neurons but not both. In contrast, groups, which consist of mixtures of both kinds of neurons, can be both. During the formation of the primary repertoire, neighboring neurons tend to connect more extensively to each other to form circuits containing varying proportions of each kind of neuron. This lends a richly cooperative property to the activity of neurons in groups, activity that one would expect to be different in different areas and maps because of differences in their primary repertoires.

There is an even more compelling reason to suppose that neuronal groups are units of selection. When maps are connected by reentrant fibers, the individual fibers generally extend their arbors over many locally linked neurons (figure 9–3). When secondary repertoires are formed, the strengthening of synapses *within* these arbors may then select groups of neighboring neurons, changing borders over smaller dimensions than those of the arbors. We may summarize by saying that, in general, no individual neuron is selected in isolation; no individual neuron in a map reenters to only one other neuron in another map; and no individual neuron has the properties alone that it shows in a group. These constraints arise from the density of the interconnections among neurons, and they make it highly unlikely that a single neuron could function as the unit of selection.

With the three tenets of the theory in hand, we are now ready to see how the ability to carry out categorization is embodied in the nervous system. I shall use the example of perceptual categorization—the selective discrimination of an object or event from other objects or events for adaptive purposes. Remember that this occurs not by classical categorization, but rather by disjunctive sampling of properties. (For a detailed discussion of the psychological bases of categorization, see the section on categories in the Postscript.)

To explain how categorization may occur, we can use the workings of what I have called a "classification couple" in the brain. This is a minimal unit consisting of two functionally different maps made up of neuronal groups and connected by reentry (figure 9–4). Each map *independently* receives signals from other brain maps or from the world (in this example, the signals come from the world). Within a certain time period, reentrant signaling strongly connects certain active combinations of neuronal groups in one map to different combinations in the other map. This occurs through the strengthening and the weakening of synapses within groups in each map and also at their connections with reentrant fibers. In this way, the functions and activities in one map are connected and correlated with those in another map. This occurs even though each map is receiving independent signals from the world: One set of inputs could be, for example, from vision, and the other from touch.

If the maps in question are topographically connected, they correlate happenings at one spatial location in the world without a higher-order supervisor. (By "topographic," I refer to the situation in which a sensory receptor sheet receiving signals from the world connects to its recipient map in such a way that neighboring locations in the sensory sheet are also neighboring locations in the recipient map.) This connectivity is not limited to a pair of maps or to any one moment of time: The interactions of *multiple* maps can be coordinated in the same fashion.

This is a very important property. Neuronal group selection occurring

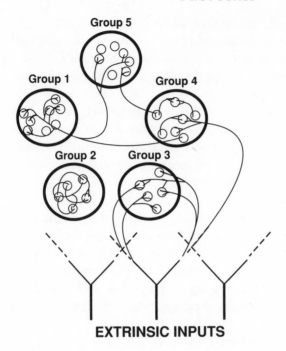

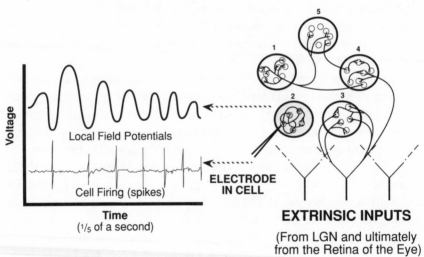

Group 5

Group 1

Group 4

Group 2

Group 3

EXTRINSIC INPUTS

Voltage

Local Field Potentials

Cell Firing (spikes)

ELECTRODE IN CELL

Time
(¹/₅ of a second)

EXTRINSIC INPUTS

(From LGN and ultimately from the Retina of the Eye)

FIGURE 9–3

Neuronal groups. Top: *How neurons are connected in groups (intrinsic connections) and how groups are connected to each other (extrinsic connections). Each group shows a different aspect of connectivity. Groups 1 and 5 show that each cell contacts cells in its own group and in other groups. Group 2 shows the dense intrinsic connectivity of groups. Group 3 shows that each group also receives inputs from a set of overlapping extrinsic inputs that can be selectively stimulated. (In general, such inputs extend over distances of many cell diameters.) Group 4 shows that each cell therefore receives inputs from cells in its own group, from cells in other groups, and from extrinsic sources. Groups differ in size* within maps leads to the production of new kinds of signals, which can then be reentered into earlier maps along with signals from the outside world. This property of reentry allows for what I have called recursive synthesis: Not only are events correlated topographically across different maps without a supervisor, but *new* selectional properties emerge through successive and recursive reentry across maps in time. This property has been simulated in a computer model, the RCI (reentrant cortical integration) model of the cerebral cortex, which I described in detail in *The Remembered Present*. This model successfully correlates the activities of many different maps by reentry and produces coordinated responses to complex visual figures.

How can reentry account for perceptual categorization, the function that the TNGS takes to be fundamental in any attempt to relate physiology to psychology? The brief answer is: By coupling the outputs of multiple maps that are reentrantly connected to the sensorimotor behavior of the animal. This is achieved through a higher-order structure called a *global mapping*. A global mapping is a dynamic structure containing multiple reentrant local maps (both motor and sensory) that are able to interact with nonmapped parts of the brain (figure 9–5). (These nonmapped parts of the brain include parts of specialized structures known as the hippocampus, the basal ganglia, and the cerebellum, the functions of which will be discussed later.) I want to stress here that a global mapping allows selectional events occurring in its *local* maps (the kind illustrated in figure 9–4) to be connected to the animal's motor behavior, to new sensory samplings of the world, and to further successive reentry events.

Such a global mapping ensures the creation of a dynamic loop that continually matches an animal's gestures and posture to the independent sampling of several kinds of sensory signals. Selection of neuronal groups within the local maps of a global mapping then results in particular categorical responses. Categorization does not occur according to a computerlike program in a sensory area which then executes a program to give a

(ranging from perhaps 50 to 10,000 neurons) and in actual connectivity, which is determined by the local neuroanatomy of the areas in which they are found. Bottom: Evidence for the existence of groups. An electrode in a visual neuron records its electrical response (spikes) as well as the responses of its neighbors (field potentials) in the group. When a visual stimulus of the right type (a lit bar moving up and to the right) is present, the responses of the neuron and its neighbors all oscillate at the same frequency (forty hertz, or forty cycles per second). When the stimulus is removed, the spikes and field potentials no longer correlate.

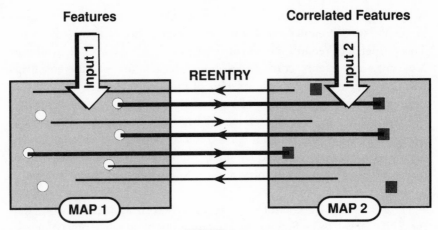

FIGURE 9–4

Reentry. Two maps of neuronal groups receive independent inputs (1 and 2). Each map is functionally segregated; that is, map 1 responds to local features (for example, visually detected angles) that are different from those to which map 2 responds (for example, an object's overall movement). The two maps are connected by nerve fibers that carry reentrant signals between them. These fibers are numerous and dense and serve to "map the maps" to each other. If within some time period the groups indicated by the circles in map 1 are reentrantly connected to the groups indicated by the squares in map 2, these connections may be strengthened. As a result of reentrant signaling, and by means of synaptic change, patterns of responses in map 1 are associated with patterns of responses in map 2 in a "classification couple." Because of synaptic change, responses to present inputs are also linked to previous patterns of responses.

particular motor output. Instead, sensorimotor activity over the whole mapping *selects* neuronal groups that give the appropriate output or behavior, resulting in categorization. Decisions in such systems are based on the statistics of signal correlations. Notice the contrast with computers; these changes occur within a selectional system rather than depending on the carriage of coded messages in a process of instruction.

But what is "appropriate" with respect to behavior, and how does perceptual categorization manifest itself? The TNGS proposes that categorization always occurs in reference to internal criteria of value and that this reference defines its appropriateness. Such value criteria do not determine specific categorizations but they constrain the domains in which they occur. According to the theory, the bases for value systems in the animals of a given species are already set by evolutionary selection. They are exhibited in those regions of the brain concerned with regulating bodily functions: heartbeat, breathing, sexual responses, feeding responses, endocrine functions, autonomic responses. Categorization manifests itself in

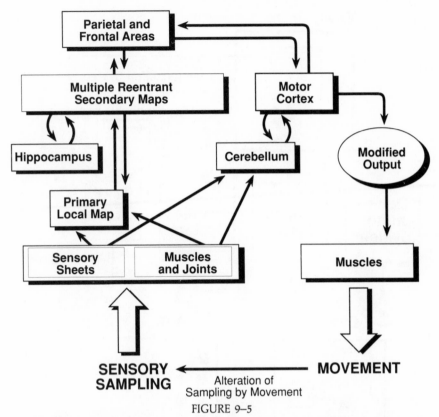

FIGURE 9–5

A global mapping. This structure is made up of multiple maps (of the kind shown in the previous figure). The maps are also connected to brain regions such as the hippocampus and cerebellum. Notice that signals from the outside world enter into this mapping, and that multiple sources of output lead to movement. This in turn alters how sensory signals are picked up. A global mapping is thus a dynamic structure, one that changes with time and behavior. Its reentrant local maps, which correlate features and movement, make perceptual categorization possible. Perturbations at different levels cause a global mapping to rearrange, to collapse, or to be replaced by another global mapping.

behavior that appropriately fulfills the evolutionarily selected requirements of such life-supporting physiological systems.

A specific example of categorization constrained by value may help connect these ideas. My colleagues and I have simulated complex automata based on the TNGS in supercomputers to demonstrate that perceptual categorization can be carried out on value in a global mapping (figure 9–6). In automata such as Darwin III, value is seen to operate for the visual system, for example, in circuits that favor light falling on the central part of the eye. (Value = "light is better than no light"; light and stimulation

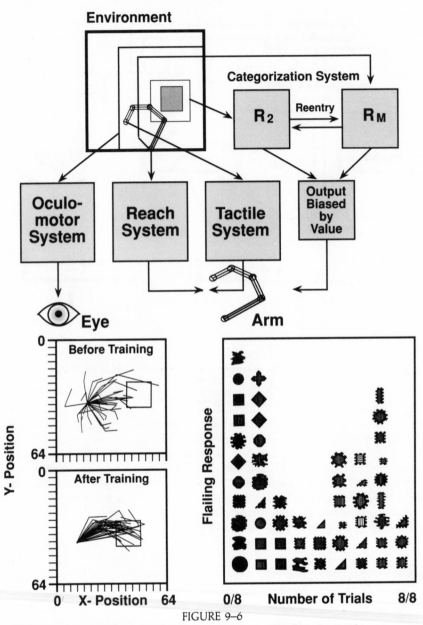

FIGURE 9–6

Darwin III, a recognition automaton that performs as a global mapping. This automaton was simulated in a supercomputer. It has a single movable eye, a four-jointed arm with touch at the last joint, and kinesthesia (joint sense) signaled by neurons in its joints as they move. Its nervous system is organized into several subsystems, each responsible for different aspects of its behavior (top). It contains many maps of the kinds shown in figures 9–4 and 9–5. What is programmed in the simulation is the "evolutionary" phenotype, including neuroanatomy. But the behavior of the simulation is not programmed (see at the center of vision are favored over light and stimulation at the periphery.) In Darwin III, the action of these value circuits enhances the probability that synapses active when such circuits are engaged will be strengthened in preference to competing synapses. The net result is that with selection and experience the eye of the automaton tracks signals from lit objects.

This defines one form of "appropriate" behavior as acquired behavior that is consistent with evolutionarily set values. But perceptual categorization occurs only when, after disjunctive sampling of signals in *several* modalities (vision, touch, joint sense), Darwin III activates an output through its reentrant maps. This occurs, for example, when as a result of explorations with its "hand-arm" and "eye" it "decides" that something is an object, that the object is striped, and that the object is bumpy. Given that Darwin III has a higher-order value system for output on such a categorical decision, it then activates a neural circuit that flails its arm. This output reflects the categorization that results from multiple synaptic selectional events that have occurred as a result of experience in all its reentrantly connected maps. These selectional events occur during ongoing behavior and are *not* specified by value. As shown in figure 9–6, Darwin III "decides" by its flailing response among a large set of objects, distinguishing those that are striped and bumpy from those that are striped or bumpy but not both. Like an animal, it does this in an individually variant manner, not by predefined classical criteria. That is, it categorizes only on the basis of experience, not on the basis of prior programming. Darwin III is a model of a global mapping that *carries out categorization on value in a fashion that might be called embodied.* In general, a global mapping such as Darwin III is the smallest organization capable of categorization. Of course, the brain of a real animal has the capacity to assemble many more mappings of this kind.

Incidently, the movements of the "arm" of Darwin III are also selected

chapter 19). After experience with randomly moving objects that it "sees," its eye will follow any object. Similarly, its arm reaches out to "touch" an object, and with each selection of movements it increases its success in achieving this touching (lower left). In the experiments shown at the lower left, the tip of the arm always starts in a standard location (the point of origin of the traces). Its motion toward a target area (the square box) is plotted. Notice that before training, the arm moves in many directions. After training involving selection (bottom set of traces), its movements are targeted. Darwin III was confronted with fifty-five different objects and was given eight trials in which to categorize each object. The results (lower right), plotted as a positive flailing response versus the number of successful trials, show that Darwin III divided this collection of objects into two classes.

rather than instructed. For example, by defining value as a factor that increases the likelihood of the strengthening of synaptic connections in those movement repertoires that are active when the automaton's arm moves to the center of the visual field, one obtains a system in which originally undirected movements eventually yield to those that successfully target an object (figure 9–6).

According to the TNGS, the driving forces of animal behavior are thus evolutionarily selected value patterns that help the brain and the body maintain the conditions necessary to continue life. These systems are called homeostats. It is the coupling of motion and sensory sampling resulting in behavior that changes the levels of homeostats. Aside from those occasional species-specific behavior patterns that have been selected for directly by evolution, however, most categorization leading to behavior that changes homeostatic levels occurs by a *somatic* selection of neuronal groups in each animal. Categorization is not the same as value, but rather occurs *on* value. It is an epigenetic developmental event, and no amount of value-based circuitry leads to its occurrence without experiential selection of neuronal groups. But it is also true that without prior value, somatic selectional systems will not converge into definite behaviors. This has been shown, for example, by cutting off the value circuits in Darwin III: Not surprisingly, the convergent behavior shown in the figures does not occur if this is done.

This has been a strenuous tour. I hope nevertheless that I have conveyed something of the flavor and self-consistency of the TNGS. Now I want to say something about the evidence that has been gathered in support of the theory since it was first proposed. I shall not be exhaustive, but because the theory has been occasionally attacked or misunderstood, it may be useful to clarify some issues and mention some experimental findings that corroborate it.

The two concepts of the theory to have come under the most intense attack are those of neuronal groups and of selection itself. Horace Barlow and, separately, Francis Crick have attacked the notion of the existence of groups. Barlow's criticism is based on the claim that neuronal group formation would require a Malthusian population dynamic. Thomas Malthus was an inspiration for Darwin, who recounts that when he read Malthus's suggestion that populations grow geometrically while food supplies grow only arithmetically, he saw how competitive struggle could lead to natural selection. But Barlow's thinking is not as clear as that of his distinguished forebear. Neither natural selection in evolution nor neuronal group selection necessarily requires a growing population. What is required is *differential* reproduction (for evolution) and *differential* amplification (for neuronal group selection by changes in synaptic strength). Barlow compounds his error by assuming the wrong mechanism of synaptic change and concluding on this basis that, during selection, neuronal groups would have to incorporate increasing numbers of cells or else lose their selectivity. Explicit models have shown that neither of these consequences results, provided one chooses rules of synaptic change that more closely resemble those found in real experiments.

Crick's claim is that neuronal groups have little evidence to support them. He also asserts that neuronal group selection is not necessary to support ideas of global mapping. Finally, he claims that he has not found it possible to make a worthwhile comparison between the theory of natural selection and what happens in the developing brain.

Contrary to these claims, experimental findings have emerged since the TNGS was first proposed that directly demonstrate the existence of neuronal groups and the functions of reentry. One of the main findings concerns the behavior of neurons in the "orientation columns" of the primary visual cortex (see figures 9–2 and 9–3). When moving bars of light are presented to an animal's visual system, particular neurons in these columns are known to fire in response to bars that are oriented in a particular direction. Different neurons have different "orientation tuning" or specific responses to bars moving at different angles. Wolf Singer and his colleagues, and Reinhard Eckhorn and his colleagues, have shown that the probability of firing of a single cell in this part of the visual cortex is very closely correlated with the firing of neighboring cells in the same column. The activity of these neighboring cells was measured by recording local field potentials, which represent the summed activity of many cells in a small area. These experiments showed that the presentation of an appropriately oriented bar caused a *group* of cells within the responding orientation column to fire together in time, with a predominant oscillatory component at forty hertz (forty cycles per second; see figure 9–3). When the stimulus was removed, the coherent oscillatory response of the group of cells disappeared. Just as strikingly, when a stimulus bar was presented and field potentials were recorded in two *separate* visual maps, V1 and V2 (see figure 9–2), the groups of neurons in the two maps showed a mutual phase-coherent oscillation at about forty hertz. That is, despite the distance between them, groups of neurons in separate maps oscillated at the same frequency and phase when the bar of light was presented. The distant neurons in these groups are known to be linked by reentrant fibers. Thus, the phasic reentrant signaling proposed in the original TNGS appears to be confirmed. My colleagues and I modeled these findings in a computer. We found that cutting even one leg of the reentrant pathway between the two areas served to make the oscillation of simulated neuronal groups in the two areas go out of phase and become incoherent.

These findings corroborate: (1) the notion of a group—cooperatively interactive neurons more or less tightly coupled by synaptic connections, firing together and responding as units to selection by particular stimuli; and (2) the concept of the correlation by reentry of selective events occurring in different maps. Both ideas have been shown to apply experimentally to secondary repertoires like those of the visual system.

We have also successfully modeled the plastic changes that occur in the map boundaries of another secondary repertoire, one located in the part of the cortex subserving touch. These changes were discovered experimentally by Michael Merzenich and his colleagues (see chapter 3, particularly the bottom part of figure 3–5). The notion of neuronal groups undergoing *competition* for interaction with neighboring neurons by strengthening their synaptic connections has nicely explained this plasticity. Changing the patterns of light touch or cutting the nerves that mediate finger touch leads to rapid changes of map boundaries in the somatosensory cerebral cortex subserving these functions. The findings are entirely compatible with a Darwinian notion of selection among groups competing within a map.

It seems that the criticisms leveled at the notion of neuronal groups do not stand up. Criticisms of the theory at the level of primary repertoire formation have also revealed deep misunderstanding. Dale Purves's interpretation that the theory is "regressive" because it suggests selection of nerve cells solely by elimination during development is simply a misrepresentation. The description of primary repertoire formation in *Neural Darwinism* explicitly states that eliminative selection is insufficient. While eliminative selection undoubtedly occurs during the formation of the nervous system, it is only one of many selectional mechanisms. Others of equal or greater importance are the formation of new anatomical paths by the expression of new adhesion molecules and the formation of signal loops by activity-dependent synapse formation.

Crick's position (see his quote at the beginning of this chapter) that the theory should not be called neural Darwinism but rather neural Edelmanism because it bears no relation to Darwin's work is simultaneously derisive and flattering. But it is also misplaced. As pointed out by Richard Michod and also by myself, the theory has definite parallels to Darwinian notions. In other words, it employs population thinking quite stringently. Corresponding to the idea of fitness, for example, a neuronal group has a likelihood of response to an input and this likelihood has to do with variant structural characteristics. The connectivity of variant groups directly affects this likelihood. Furthermore, there is a relationship between the idea of heritability and neuronal group selection. In a selective system, there must be some correlation higher than background noise between parent and offspring entities. In evolution this is assured by inheritance, and in the TNGS by synaptic change. Neuronal groups that respond initially to a stimulus have, on the average, a higher likelihood of responding to a similar stimulus when it is subsequently presented, but that likelihood is modulated by value systems. In evolution, differences among various organisms' adaptations to the environment lead to differences among reproductive processes, which lead in turn to changes in the frequencies of genes in the population. In neuronal group selection, differences in connectivity, synaptic structure, and the morphology of neurons in the primary repertoire, after confrontation with different correlated patterns of signals from the environment, lead to differences in the probabilities of their responses as groups. This reflects changes in the patterns of their synaptic strengths. There is differential reproduction in one case, differential amplification in the other. Crick's misprision is probably based on his faulty notion that the elements in a repertoire cannot vary as a result of selection and that therefore they must be absolutely fixed entities. This is true neither for natural selection nor for neuronal group selection.

It is crucial to recognize that while the *principles* of the sciences of recognition (evolution, immunity, and brain science) are shared, their *mechanisms* must obviously be different. What is stunning about these sciences is that *natural selection* during evolution produced two completely different *somatic selection* systems capable of recognition. If we accept these ideas, a small loop consisting of the events of neuronal group selection leads to diverse phenotypic behaviors in different individuals of a species. These diverse behaviors provide the basis for ongoing natural selection in the grand loop of evolution. The two selective systems, somatic and evolutionary, interact.

Everything in scientific inquiry should be exposed to remorseless criticism. What is curious about the criticisms of the TNGS is the level at which they have been aimed. One would have expected that most criticism would have been aimed at the attempt to bridge psychology and physiology—that is, at the proposed mechanisms of perceptual categorization and memory. These mechanisms, along with the proposals for concept formation discussed later, are at the true heart of the theory, and they remain to be tested. As matters stand now, however, neither the experimental findings on which the TNGS is based nor the actual proposals of the theory itself have been displaced. Indeed, several predictions of the theory have already been confirmed. It would be enormously valuable if either the facts I have presented were shown by scientists to be false or if an alternative theory based on them were forthcoming. We could then look forward to more constructive criticisms and to developments that might further sharpen our vision of brain function.

So far this has not been the case and accordingly I now turn to those parts of the theory concerned with bridging the gap between physiology and psychology. It is across that bridge that a biological account of consciousness must pass.

CHAPTER 10

Memory and Concepts: Building a Bridge to Consciousness

'Concept' is a vague concept.

—Ludwig Wittgenstein

What is an idea?
  It is an image that paints itself in my brain.

—Voltaire

This is a good point at which to take stock and to look ahead. We have been trying to see how mental functions are embodied, how psychology maps onto physiology.

We have argued that natural selection has given rise to somatic selectional systems—the immune system and the brain. The major basis of brain function is morphology. The appropriate neuroanatomy develops according to topobiological principles. Indeed, the brain is a topobiological system par excellence, consisting as it does of maps and mapping systems in which place is critical for performance.

Two apparently unrelated observations have compelled us to take a new look at how the brain might function as a recognition system. The first is the enormous diversity and individuality of brain structure. The second is that the world, although constrained by the laws of physics, is an unlabeled place. To consider how a brain so constituted might categorize such a world, the TNGS was formulated. Its tenets—developmental selection, experiential selection, and reentry—are considered to be the fundamental grounds for developing psychological functions. This does not mean, however, that new morphological arrangements are not necessary for *emergent* psychological functions. It just means that the TNGS assumes that no additional major principles have to be added to assure the evolution of new functions. I will argue here that somatic selection acting in global mappings, with new kinds of mappings added to old ones during evolution, is a powerful means of acquiring new functions such as specialized memories and conceptual abilities.

Before turning to a consideration of memory and concepts themselves, it may be useful to look at how we expect "higher brain functions" to be related. What psychological functions should this selectionist view explain? And how can they account for consciousness and intentionality?

The fundamental triad of higher brain functions is composed of perceptual categorization, memory, and learning. (While these functions are often treated separately for the convenience of discussion, it should be kept in mind that, in fact, they are inseparable aspects of a common mental performance.) We have already seen how classification couples and global mapping can carry out perceptual categorization. Perceptual categorization is generally necessary for memory, which is, after all, about previous categorizations. The functioning of both may be tested by analyzing behavior. We will see in this chapter that any kind of memory, while based on changes in synaptic strength, is a dynamic system property, one whose characteristics depend on the actual neural structures in which it occurs. To serve the adaptive needs of an animal faced with the unforeseen juxtapositions of events affecting survival, however, learning that affects behavior is also necessary. Thus, the three fundamental functions—categorization, memory, and learning—are closely connected: The last depends on the first two.

Yet while perceptual categorization and memory are necessary for learning, they are not sufficient. What is needed in addition is a connection to value systems mediated by parts of the brain that are different from those that carry out categorization. The sufficient condition for adaptation is provided by the linkage of global mappings to the activity of the so-called hedonic centers and the limbic system of the brain in a way that satisfies homeostatic, appetitive, and consummatory needs reflecting evolutionarily established values. These value-laden brain structures, such as the hypothalamus, various nuclei in the midbrain, and so on, evolved in response to ethological demands, and some of their circuits are species-specific. It is obvious why this is so: Mating activity and behavior in birds varies widely from that in whales.

Learning in any species results from the operation of neural linkages between global mappings and the value centers mentioned above. It serves to connect categorization to behaviors having adaptive value under conditions of expectancy. Physiological systems, like some control devices, have set points (think of a thermostat). What is meant by expectancy is simply the condition under which the set points of the physiological structures making up portions of the hedonic system are not yet satisfied. Learning is achieved when behavior leads to synaptic changes in global mappings that satisfy the set points.

We can now see why the operation of memory, as it relates perceptual categorization to learning, strongly underlines the adaptive value of neuronal group selection. Increasing the size of the primary repertoires or the reentrant connectivity between repertoires, or enhancing the means of synaptic change by adding new chemical mechanisms during evolution, increases the number of categorical responses that may enhance learning. Learning is adaptive, and by this reasoning, having more numerous or more diversified neuronal groups would also be adaptive. Nevertheless, whatever the degree of learning, behavior is constrained by ethological factors, among the most important of which are the value systems and homeostatic requirements selected for during the evolution of a species.

Memory is at the center of all these events, and I shall spend a good part of the rest of this chapter analyzing its workings and requirements. As important as the basic triad of perception, memory, and learning is, however, their functioning together cannot generate the kinds of capabilities that connect perceptual categorizations together to yield general *relational* properties. These properties emerge from the acquisition of conceptual capabilities—the ability to categorize in terms of general or abstract relations. So I shall also have to discuss the subject of concepts. My goal is to show how, with the physiological bases for our central triad and for conceptual capabilities in place, we can account for the emergence of consciousness without invoking any new principles beyond those already contained in the TNGS.

MEMORY

Let us begin with memory. One difficulty in dealing with memory is that so many different kinds have been described, and so many of them are so closely related to linguistic capability that it becomes difficult to tease out the fundamentals. As a result, a physiological basis for memory—synaptic change—is often mistakenly equated with memory itself in an attempt to simplify matters.

To clarify the issue, let us agree that, whatever form it takes, memory is the ability to repeat a performance. The kind of performance depends on the structure of the system in which the memory is manifest, for memory is a system property. As such, memory in the nervous system is a dynamic property of populations of neuronal groups. In computers, memory depends on the specification and storage of bits of coded information. According to the TNGS, this is *not* the case in the nervous system. For example, while the behavior of Darwin III obviously shows signs of a form of memory after its encounters with various objects (see the last chapter, especially figure 9–6), its memory cannot be described in terms of a static configuration of bits. The memory exhibited by a global mapping such as Darwin III is not a store of fixed or coded attributes that can be called up and assembled in a replicative fashion, as is the case in a computer.

The TNGS proposes instead that memory is the specific enhancement of a previously established ability to categorize (figure 10–1). This kind of memory emerges as a population property from continual dynamic changes in the synaptic populations within global mappings—changes that allow a categorization to occur in the first place. Alterations in the synaptic strengths of groups in a global mapping provide the biochemical basis of memory.

In such a system, recall is not stereotypic. Under the influence of continually changing contexts, it changes, as the structure and dynamics of the neural populations involved in the original categorization also change. Recall involves the activation of some, but not necessarily all, of the previously facilitated portions of global mappings. It can result in a categorization response similar to a previous one, but at different times the elements contributing to that response are different, and in general they are likely to have been altered by ongoing behavior.

Since perceptual categories are not immutable and are altered by the ongoing behavior of the animal, memory, in this view, results from a process of continual *recategorization*. By its nature, memory is procedural and involves continual motor activity and repeated rehearsal in different contexts. Because of the new associations arising in these contexts, because of changing inputs and stimuli, and because different combinations of neuronal groups can give rise to a similar output, a given categorical response in memory may be achieved in several ways. Unlike computer-based memory, brain-based memory is inexact, but it is also capable of great degrees of generalization.

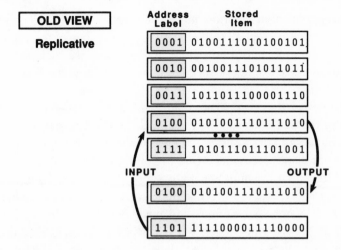

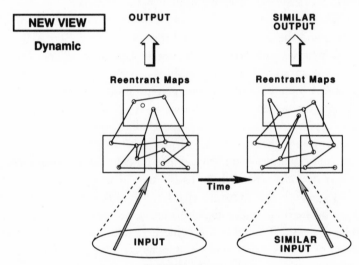

FIGURE 10–1

Two views of memory. Top: An example of memory as the storage of precisely coded information (replicative memory). I call it replicative because recall must reproduce the same coded pattern without error and thus replicate it faithfully as in a computer. One change in a bit anywhere is an error. Bottom: An example of dynamic memory in a global mapping of the kind illustrated by Darwin III (figure 9–6) after it carries out categorization on value. Many similarly categorized objects can give the same output, and mistakes can be made. This memory is a property of the entire system, although its fundamental mechanism is change in synaptic strength, as indicated by changes in the lines between the neuronal groups (small circles) inside the maps.

The properties of association, inexactness, and generalization all derive from the fact that perceptual categorization, which is one of the initial bases of memory, is probabilistic in nature. It is no surprise that different individuals have such different memories and that they use them in such different fashions.

Writing about these properties, I am reminded of the contrast between different gifted individuals in their approach to memory and performance. A story told about Fritz Kreisler, the great violinist, and Sergei Rachmaninoff, the great pianist, provides a case in point. In 1930, they met in Berlin to record the Grieg C Minor Sonata together. Rachmaninoff was a meticulous worker and wanted to practice right away. Kreisler, who didn't practice much, was not so assiduous and went out on the town. The next morning, at Kreisler's insistence and with Rachmaninoff's reluctance, the recording took place; it went well. (It is still available, I believe, and is stunning.) Nonetheless, Rachmaninoff was not pleased.

Somewhat later that year, the story goes, the two played together in New York, and the program included the same sonata. Somewhere in the course of a movement, Kreisler had a memory lapse. Being Kreisler, he simply made up some cadences, probably in the hope of picking up the thread later. When that didn't happen after a minute or so, he leaned over, still playing, and asked, "Sergei, where are we?" Rachmaninoff looked up from the keyboard and said, "Carnegie Hall."

If one considers memory to be a form of recategorization, it is obvious that one can only understand its workings by considering the entire system in which it operates. (Refer back to Darwin III in the last chapter for an example.) One of the dynamic characteristics of the system of global mappings in the brain is the ability to order successive changes. Memory would be useless if it could not in some way take account of the temporal succession of events—of sensory events as well as patterns of movement.

To see how all this works would immerse us in a sea of technical details. But it is valuable to understand a bit about the means by which the cerebral cortex and its appendages deal with the time and space requirements of memory. Remember that the cortex is an interconnected six-layered sheet of about ten billion neurons with about a million billion connections. Besides being arranged in functionally segregated maps that are reentrantly connected and that subserve all the different sensory modalities and motor responses, the cortex is connected to three structures I have called the organs of succession, obviously because they have to do with ordering the output of the brain.

Each of these structures—the cerebellum, the hippocampus, and the basal ganglia—is concerned with a different aspect and scale of ordering (figure 10–2). The cerebellum is a remarkable structure surrounding the upper brain stem. It consists of a distinctive set of neural circuits with a rather stereotyped structure, and receives two main kinds of inputs from the cerebral cortex and spinal cord. A variety of studies suggests that while the cerebellum is not absolutely required for the initiation of movement, it plays a very specific role in the timing and smoothing of successions of movements. Together with the cerebral cortex, it provides the basis for producing and categorizing smooth gestures. In the absence of portions of the cerebellum, otherwise smooth patterns of movement become jerky and discoordinated.

But what has this to do with memory? Remember that categorization depends on smooth gestures and postures as much as it does on sensory sheets. The cerebellum and motor cortex together undergo the synaptic changes yielding the smooth movements that underlie both categorization and recategorization.

The longer-range execution of a sequence of motor events, called a motor program, depends on another set of cortical appendages, the basal

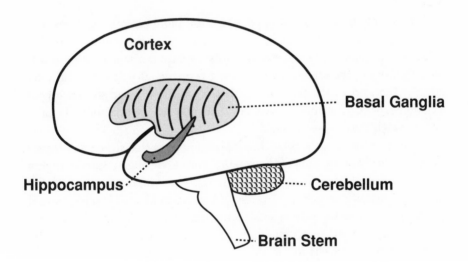

CORTICAL APPENDAGES
FIGURE 10–2

Cortical appendages—the organs of succession. The brain contains structures such as the cerebellum, the basal ganglia, and the hippocampus that are concerned with timing, succession in movement, and the establishment of memory. They are closely connected with the cerebral cortex as it carries out categorization and correlation of the kind performed by global mappings (see figure 9–5). The diagram is simply to help the reader locate these "organs of succession" in a cartoon of the brain.

ganglia. These are a large and complex set of structures located deep at the center of the brain. They connect to the cerebral cortex in a series of parallel circuits involved in eye and body movements, as well as to the frontal portions of the cortex, the function of which is related to behavioral planning and emotions. It appears that the basal ganglia are involved in planning for movement and thus in choices of the types and successions of motor output. They not only help regulate movement in a motor program by coupling sensory and motor responses but also help direct what is to be done according to a motor plan. Notice that, unlike the cerebellum, which smooths and coordinates more immediate gestures, this appendage works over longer time scales and helps correlate whole sequences of gestures in a plan.

The central role of the basal ganglia in motion and motor plans is seen in disease. In Parkinson's disease, for example, destruction of a particular set of basal ganglia (substantia nigra) neurons that produce the neurotransmitter dopamine leads to difficulty in initiating motions, to tremors, and to altered gait. The basal ganglia are also intimately connected to the hedonic centers of the brain, and as I discuss later, they very likely play a role in attention.

A third important cortical appendage, the hippocampus, also has intimate connections with the hedonic centers found in the midbrain and the hypothalamus. Its main characteristic is its important role in relating short-term memory to the establishment of long-term memory. It sits at the inner edge of the skirt of the temporal cortex, a sausage-shaped structure with a double-nested, C-shaped cross section (its appearance prompted its Latin name, which means "sea horse").

What is extraordinary about the hippocampus is that it receives inputs from practically *all* regions of the cerebral cortex, through a smaller region known as the entorhinal cortex (figure 10–2). These inputs run through the hippocampus in a sequence of three successive synapses. Having passed through these structures, the signals loop back to the entorhinal cortex and are relayed back by reentrant fibers to the cortical areas that originally connected with it. Cells inside the hippocampal loop all receive simultaneous connections indirectly from the midbrain and hedonic areas, areas subserving value.

What is all this circuitry for? It appears that the hippocampus is necessary for the laying down of long-term memories. The famous patient H. M., whose hippocampus was removed because of potentially lethal epilepsy, had a clear-cut syndrome after surgery. He could recall all long-term memories up to a time somewhat before the removal of his hippocampus. But he was unable to remember events that had occurred just a short time before he was asked about them. If interrupted in the recital of his address, for example, he lost the capacity to recall it. He was fully conscious and could even learn some motor sequences. He was, however, permanently crippled and could not store any new ongoing long-term memories because of his hippocampal defect.

Studies with animals have confirmed the role of the hippocampus in transforming responses to short-term tasks for storage into long-term memory. It appears that the role of this cortical appendage is to help order events that have been immediately categorized by the cortex and then ensure that these categorized events effect further synaptic changes in the cortex to enable long-term memory.

What all this means in terms of the TNGS is that while classification couples and global mappings undergo neuronal group selection by synaptic change, this in itself is not enough to assure the relationship between short-term and long-term memory. Unless organs of succession such as the hippocampus intervene and order the results, it appears that severe memory defects will ensue.

I have had to burden this discussion with some of the anatomical and physiological details to emphasize the interactive role of different brain structures and brain dynamics in carrying out psychological functions. Had the structures and neuroanatomy of the cerebellum, or something like it, not evolved, smoothly coordinated and rapid motion would be compromised. Without the basal ganglia and their specific anatomy, animals would not be able to orchestrate whole symphonies of movements in a plan. Without the functions provided by the hippocampus, whole suites of categorization in a time range between the immediate and those forever stored could not be linked. And without that linkage, no long-term memory could be coherent.

In these examples, we see that evolutionarily developed brain morphology and circuitry, modulated by biochemistry at the synapses, can yield new functions and new kinds of memory. It is clear from studies of invertebrates and "lower" vertebrates that the nervous system acts to regulate bodily functions and behavior. Particular types of memory based on synaptic change certainly occur in such animals. But in terrestrial life (and for the precursors of hominids, in arboreal life), great and novel environmental challenges occurred. The further evolution of the cortex for carrying out perceptual categorizations and of the organs of succession for ordering these categorizations allowed for much richer sets of psychological functions with which to deal with complex environments. These developments altered the meaning of what it is to have a memory. Notice, however, that no new principles beyond selection and reentry are necessary to gain new memory functions. What is needed are new structures or morphologies—new orderings of connections in the brain such as those seen in the cortical appendages.

CONCEPTS

The same principle applies to categorization itself. Indeed, the embodiment of broad categorical capabilities requires another evolutionary development in addition to memory as recategorization. I have called this the ability to have concepts, and the view I have taken, like the one I have taken of memory, departs from the conventional picture. The word "concept" is generally used in connection with language, and is used in contexts in which one may talk of truth or falsehood. I have used the word concept, however, to refer to a capability that appears in evolution prior to the acquisition of linguistic primitives. What is this capability?

An animal capable of having concepts identifies a thing or an action and on the basis of that identification controls its behavior in a more or less general way. This recognition must be relational: It must be able to connect one perceptual categorization to another, apparently unrelated one, even in the absence of the stimuli that triggered those categorizations. The relations that are captured must allow responses to general properties—"object," "up-down," "inside," and so on. Unlike elements of speech, however, concepts are *not* conventional or arbitrary, do *not* require linkage to a speech community to develop, and do *not* depend on sequential presentation. Conceptual capabilities develop in evolution well before speech. Although they depend on perception and memory, they are *constructed* by the brain from elements that contribute to both of these functions.

It is difficult to know which animals beside humans have conceptual abilities. Certainly the evidence on chimpanzees is persuasive. These animals generalize and classify relations—whether of things or of actions. Decisions about the conceptual capabilities of other animals are harder to make, however, because unlike the case with the chimpanzee, our communication with other animals is severely restricted. The best we may be able to do is to compare the structures and functions of their brain regions with those of humans and make guesses to guide further study.

How did conceptual abilities arise? The TNGS proposes that the evolutionary development of specialized brain areas is required before conceptual abilities emerge. The argument supporting this proposal is based on the notion that a simple increase in the number of reentrant maps capable of perceptual categorization is insufficient to account for concepts. Conceptual categorizations are enormously heterogeneous and general. Concepts involve mixtures of relations concerning the real world, memories, and past behavior. Unlike the brain areas mediating perceptions, those mediating concepts must be able to operate without immediate input.

What brain operations give rise to these properties? The TNGS suggests that in forming concepts, the brain constructs maps of its *own* activities, not just of external stimuli, as in perception. According to the theory, the brain areas responsible for concept formation contain structures that categorize, discriminate, and recombine the various brain activities occurring *in different kinds of global mappings*. Such structures in the brain, instead of categorizing outside inputs from sensory modalities, categorize parts of past global mappings according to modality, the presence or absence of movement, and the presence or absence of relationships between perceptual categorizations (figure 10–2).

Structures able to perform these activities are likely to be found in the frontal, temporal, and parietal cortices of the brain. They must represent *a mapping of types of maps*. Indeed, they must be able to activate or reconstruct portions of *past* activities of global mappings of different types—for example, those involving different sensory modalities. They must also be able to recombine or compare them. This means that special reentrant connections from these higher-order cortical areas to other cortical areas and to the hippocampus and basal ganglia must exist to carry out concepts.

Brain areas giving rise to concepts must be able not only to stimulate parts of past global mappings but also to do so independently of current sensory input. They must also be able to distinguish classes of global mappings (for instance, those corresponding to objects from those corresponding to movements). They must then be able to connect reactivated portions of global mappings and mediate the long-term storage of such activities. This is necessary because concept formation requires memory.

The frontal cortex is a prime example of a conceptual center in the brain. Not enough is known about how its maps are organized to be sure whether concept formation in this cortical area requires *topographic* mapping, as perceptual categorization does. It seems likely, however, that maps that map the types of activity occurring in other cortical maps would be required; in higher-order maps, topography may not be so important. Given its connections to the basal ganglia and the limbic system, including the hippocampus, the frontal cortex also establishes relations subserving the categorization of values and sensory experiences themselves. In this way, conceptual memories are affected by values—an important characteristic in enhancing survival.

With this notion of concepts, in which the brain categorizes its own activities (particularly its perceptual categorizations), it becomes possible to see how generalized categories and images might be embodied. It is also possible to see how events may be categorized as "past" without necessitating their being played out in present brain activities, as they must be for short-term memory and for the hippocampal succession leading to long-term memory. Furthermore, one can see how concept areas, by recursively restimulating portions of global mappings containing previous synaptic changes, give rise to *combinations* of relationships and categories. There is no need for any inherent logical order, classical categorization, or prior explicit programming. Yet the means of concept formation described here could quite naturally be responsible for establishing the complex categories that I take up in the Postscript. Finally, because concept formation is based on the central triad of perceptual categorization, memory, and learning, it is, by its very nature, intentional.

This discussion of memory as recategorization, and of concepts as the products of the brain categorizing its own activities, provides the bridging elements required for reaching our goal: a biological account of consciousness. Building on the tenets of the TNGS, the fundamental triad of perceptual categorization, memory, and learning were linked to the emergence of conceptual capabilities. Notice, however, that no new theoretical assumptions were made, only assumptions about evolutionary changes in cortical morphology that alter the patterns of reentrant connectivity. It will turn out that an additional alteration of reentrant connectivity also provides a key to understanding how we came to be aware.

CHAPTER 11

Consciousness:
The Remembered Present

Something definite happens when to a certain brain-state a certain 'sciousness' corresponds.
—William James

Most people, asked what it is about the mind that is truly distinctive and strange, would probably hark back to Descartes' lonely music of the self and say, "Consciousness." We are now at that point in our excursion when we may profitably ask whether we can do better than postulate a thinking substance that is beyond the reach of a science of extended things.

What is daunting about consciousness is that it does not seem to be a matter of behavior. It just *is*—winking on with the light, multiple and simultaneous in its modes and objects, ineluctably ours. It is a process and one that is hard to score. We know what it is for ourselves but can only judge its existence in others by inductive inference. As James put it, it is something the meaning of which "we know as long as no one asks us to define it."

Indeed, it is initially best defined by considering some of its properties (of course the temptation is to indulge in a circular definition, made in terms of "awareness"). Consider what I call its "Jamesian" properties (after James, who discussed them): It is personal (possessed by individuals or selves); it is changing, yet continuous; it deals with objects independent of itself; and it is selective in time, that is, it does not exhaust all aspects of the objects with which it deals.

Consciousness shows intentionality; it is of or about things or events. It is also to some extent bound up with volition. Some psychologists suggest that consciousness is marked by the presence of mental images and by their use to regulate behavior. But it is *not* a simple copy of experience (a "mirror of reality"), nor is it necessary for a good deal of behavior. Some kinds of learning, conceptual processes, and even some forms of inference proceed without it.

I have made a distinction, which I believe is a fundamental one, between primary consciousness and higher-order consciousness. Primary consciousness is the state of being mentally aware of things in the world—of having mental images in the present. But it is not accompanied by any sense of a person with a past and future. It is what one may presume to be possessed by some nonlinguistic and nonsemantic animals (which ones they may be, I discuss later on). In contrast, higher-order consciousness involves the recognition by a thinking subject of his or her own acts or affections. It embodies a model of the personal, and of the past and the future as well as the present. It exhibits direct awareness—the noninferential or immediate awareness of mental episodes without the involvement of sense organs or receptors. It is what we as humans have in addition to primary consciousness. We are conscious of being conscious.

There are other resonances in the term "consciousness"—these are revealed, for example, in the criteria used by clinicians to assess whether a traumatized patient is "conscious" or not—criteria concerned with alertness, orientation, self-awareness, and motivational control. Physicians talk of consciousness as being "clouded," in which state perceptual acuity and memory capacity are diminished. In extreme cases of disease, the Jamesian properties, the "flights and perchings of consciousness," become random, automatized, or show perseveration, with no evidence of the existence of introspection or any attention to novelty. And in the last extreme— nothing, nothing to report.

There is no end of hypotheses about consciousness, particularly by philosophers. But most of these are not what we might call principled scientific theories, based on observables and related to the functions of the brain and body. Several theories of consciousness based on functionalism and on the machine model of the mind (see the Postscript) have recently been proposed. These generally come in two flavors: one in which consciousness is assumed to be efficacious, and another in which it is considered an epiphenomenon. In the first, consciousness is likened to the executive in a computer systems program, and in the second, to a fascinating but more or less useless by-product of computation.

In none of these notions, however, is there a direct appeal to biology or to the nature of embodiment. Such an appeal would obviously be essential to any theory of consciousness that is based on evolution. A theory of this kind must propose explicit *neural* models that explain how consciousness arises. It must of necessity explain how consciousness emerges during evolution and development. It must connect consciousness to other mental matters such as concept formation, memory, and language. And it must describe stringent tests for the models it proposes in terms of neurobiological facts. These tests should be undertaken, preferably with real experiments, or at least with what are called *gedankenexperiments*—thought experiments. In the latter, any properties postulated must be completely consistent with presently known scientific observations from whatever field of inquiry and, above all, with those from brain science.

Given the present state of affairs, this is a tall order because analyses of consciousness in biology are a bit like analyses of early cosmological events: Right from the beginning, certain manipulations and observations are just not possible. Under these circumstances, one must be careful to spell out the assumptions underlying any proposed theory. I will spell out three that are part of the underpinnings of my theory of consciousness. Two of these assumptions are straightforward, but the third is a bit tricky. I call them the physics assumption, the evolutionary assumption, and the qualia assumption (the tricky one). I have to make these assumptions clear beforehand to avoid certain pitfalls, for example, into the Cartesian position, into panpsychism, or into the cognitivist–objectivist quagmire that I discuss in the Postscript.

The physics assumption is that the laws of physics are not violated, that spirits and ghosts are out; I assume that the description of the world by modern physics is an adequate but not completely sufficient basis for a theory of consciousness. Modern quantum field theory provides a description of a set of formal properties of matter and energy at all scales (see figure P–1). It does not, however, include a theory of intentionality or a theory of names for macroscopic objects, nor does it need them. What I mean by physics being just adequate is that I allow no spooks—no quantum gravity, no action at a distance, no superphysics (see the Postscript)—to enter into this theory of consciousness.

The evolutionary assumption is also reasonably straightforward. It is that consciousness arose as a phenotypic property at some point in the evolution of species. Before then it did not exist. This assumption implies that the acquisition of consciousness either conferred evolutionary fitness directly on the individuals having it, or provided a basis for other traits that enhanced fitness. The evolutionary assumption implies that consciousness is *efficacious*—that it is *not* an epiphenomenon ("merely the redness of the melting metal," when pouring is what counts).

Now, however, with the third assumption, we come to more subtle issues. They are methodological ones, forced on us by the peculiar way in which consciousness is made manifest. To explain the difficulty, I must make a detour here to discuss phenomenal or felt properties, otherwise known as qualia.

Qualia constitute the collection of personal or subjective experiences, feelings, and sensations that accompany awareness. They are phenomenal states—"how things seem to us" as human beings. For example, the "redness" of a red object is a quale. Qualia are discriminable parts of a mental scene that nonetheless has an overall unity. They may range in intensity and clarity from "raw feels" to highly refined discriminanda. These sensations may be very precise when they accompany perceptual experiences; in the absence of perception, they may be more or less diffuse but nonetheless discernible as "visual," "auditory," and so on. In general, in the normal waking state, qualia are accompanied by a sense of spatiotemporal continuity. Often, the phenomenal scene is accompanied by feelings or emotions, however faint. Yet the actual *sequence* of qualia is highly individual, resting on a series of occasions in one's own personal history or immediate experience.

Given the fact that qualia are experienced directly only by single individuals, our methodological difficulty becomes obvious. *We cannot construct a phenomenal psychology that can be shared in the same way as a physics can be shared.* What is directly experienced as qualia by one individual cannot be fully shared by another individual as an observer. An individual can report his or her experience to an observer, but that report must always be partial, imprecise, and relative to his or her own personal context. Not only are qualia fleeting, but interventions designed to probe them may change them in unforeseen ways. Furthermore, many conscious and nonconscious processes simultaneously affect each person's subjective experience. Individuals may have their own private theories of the totality of their *individual* conscious experiences, but these can never be scientific theories. This is because other observers do not have adequate experimental controls available to them.

The paradox is a poignant one: To do physics, I employ my conscious life, perceptions, and qualia. But in my intersubjective communication, I leave them out of my description, assured that fellow observers with their own individual conscious lives can carry out the prescribed manipulations and achieve comparable experimental results. When for some reason qualia do affect interpretations, the experimental design is modified to exclude such effects; the mind is removed from nature.

But in investigating consciousness, we cannot ignore qualia. The dilemma is that phenomenal experience is a first-person matter, and this

114

seems, at first glance, to prevent the formulation of a completely objective or causal account. Is this a completely hopeless situation?

I think not. But what alternatives are open to us if we want to pursue a scientific analysis of consciousness? One alternative that definitely does not seem feasible is to ignore completely the reality of qualia, formulating a theory of consciousness that aims *by its descriptions alone* to convey to a hypothetical "qualia-free" observer what it is to feel warmth, see green, and so on. In other words, this is an attempt to propose a theory based on a kind of God's-eye view of consciousness. But no *scientific* theory of whatever kind can be presented without already assuming that observers have sensation as well as perception. To assume otherwise is to indulge the errors of theories that attempt syntactical formulations mapped onto objectivist interpretations—theories that ignore embodiment as a source of meaning (see the Postscript). There is no qualia-free scientific observer.

If we exclude such an avenue, what other recourse is there? I believe there is one, based on the fact that human beings are in a privileged position. While we may not be the only conscious animals, we are, with the possible exception of the chimpanzee, the only self-conscious animals. We are the only animals capable of language, able to model the world free of the present, able to report on, study, and correlate our phenomenal states with the findings of physics and biology.

This suggests an approach to the problem of qualia. As a basis for a theory of consciousness, it is sensible to *assume* that, just as in ourselves, qualia exist in other conscious human beings, whether they are considered as scientific observers or as subjects. (It does not matter whether these qualia are exactly the same in all observers, only that they exist.) We can then take human beings to be the best canonical referent for the study of consciousness. This is justified by the fact that human subjective reports (including those about qualia), actions, and brain structures and function *can all be correlated.* After building a theory based on the assumption that qualia exist in human beings, we can then look anew at some of the properties of qualia based on these correlations. It is our ability to report and correlate while individually experiencing qualia that opens up the possibility of a scientific investigation of consciousness.

This *qualia assumption* distinguishes between higher-order consciousness and primary consciousness. Higher-order consciousness is based on the occurrence of direct awareness in a human being who has language and a reportable subjective life. Primary consciousness may be composed of phenomenal experiences such as mental images, but it is bound to a time around the measurable present, lacks concepts of self, past, and future, and lies beyond direct descriptive individual report from its own standpoint.

Accordingly, beings with primary consciousness alone cannot construct theories of consciousness—even wrong ones!

A research program built on the assumptions I have discussed obviously has a number of inherent difficulties. We must first build a model for primary consciousness, build on that a model for higher-order consciousness, and then proceed to check the connections of each of these models with human phenomenal experience. To be consistent with the evolutionary assumption, this procedure must explain how primary consciousness evolved, and then explain how it was followed by higher-order consciousness. The order of the experimental enterprise (which, according to the qualia assumption, must be based on correlations obtained mainly on human subjects) must therefore be exactly opposite that of the theoretical one, which must begin with the evolutionary precursors to humans.

I hope it is now clear why a biological theory based on our three assumptions cannot take a God's-eye view. To be scientists, we cannot expect any theory of consciousness to render obvious to a hypothetical qualia-free animal what qualia are by any linguistic description. To maintain intersubjective communication and carry out scientific correlation, which is a human activity, we *must* assume qualia. Qualia cannot be derived as experiences from any theory. This does not mean, however, that different qualia cannot be theoretically discriminated in terms of modality, intensity, continuity, or their temporal and spatial properties. Nor does it mean that, after making the qualia assumption, we cannot consider the actual mechanisms by which qualia arise. Our cosmological comparison is not so far afield; we may ask modern physics to explain certain aspects of cosmology beginning at the earliest moment, consistent with the understanding given to us by modern physical theory. But we cannot ask a theory of physics to give a satisfactory answer to Gottfried Leibniz's question of why there is something rather than nothing.

As it will turn out after we consider models for primary and higher-order consciousness, qualia may be usefully viewed as forms of higher-order categorization, as relations reportable to the self and then somewhat less satisfactorily reportable to others with similar mental equipment. Such a terse statement hardly satisfies. But instead of expanding on it now, I will describe a model of primary consciousness, based on our three assumptions, that appears to be consistent with the facts of brain structure and function. The elements of this model include several systems already discussed, ones that give rise to value, to perceptual and conceptual categorization, and to memory. The dynamics of the model depend on a special kind of reentrant circuit. This is why I have explained these matters at length in previous chapters. (I will keep qualia to one side for now, but I will return to them later when considering higher-order consciousness.)

PRIMARY CONSCIOUSNESS

The model I have proposed has a number of parts. (Would you believe a model of consciousness that had only one part?) Before describing their interactions, I want to say a few things about each part that might make an explanation of their interactions clearer. There are, grossly speaking, two kinds of nervous system organization that are important to understanding how consciousness evolved. These systems are very different in their organization, even though they are both made up of neurons. The first is the brain stem, together with the limbic (hedonic) system, the system concerned with appetite, sexual and consummatory behavior, and evolved defensive behavior patterns. It is a value system; it is extensively connected to many different body organs, the endocrine system, and the autonomic nervous system. Together, these systems regulate heart and respiratory rate, sweating, digestive functions, and the like, as well as bodily cycles related to sleep and sex. It will come as no surprise to learn that the circuits in this limbic–brain stem system are often arranged in loops, that they respond relatively slowly (in periods ranging from seconds to months), and that they do not consist of detailed maps. They have been selected during evolution to match the body, not to match large numbers of unanticipated signals from the outside world. These systems evolved early to take care of bodily functions; they are systems of the interior.

The second major nervous system organization is quite different. It is called the thalamocortical system. (The thalamus, a central brain structure, consists of many nuclei that connect sensory and other brain signals to the cortex.) The thalamocortical system consists of the thalamus and the cortex acting together, a system that evolved to receive signals from sensory receptor sheets and to give signals to voluntary muscles. It is very fast in its responses (taking from milliseconds to seconds), although its synaptic connections undergo some changes that last a lifetime. As we have seen, its main structure, the cerebral cortex, is arranged in a set of maps, which receive inputs from the outside world via the thalamus. Unlike the limbic–brain stem system, it does not contain loops so much as highly connected, layered local structures with massively reentrant connections. In many places these are topographically arranged (see figure 9–2). The cerebral cortex is a structure adapted to receive a dense and rapid series of signals from the world through many sensory modalities simultaneously—sight, touch, taste, smell, hearing, joint sense (feeling the position of your extremities). It evolved later than the limbic–brain stem system to permit increasingly sophisticated motor behavior and the categorization of world events. To handle time as well as space, the cortical appendages—the cerebellum, basal ganglia, and hippocampus (see figure 10–2)—evolved along with the cortex to deal with succession both in actual motion and in memory.

The two systems, limbic–brain stem and thalamocortical, were linked during evolution. The later-evolving cortical system served learning behavior that was adaptive to increasingly complex environments. Because this behavior was clearly selected to serve the physiological needs and values mediated by the earlier limbic–brain stem system, the two systems had to be connected in such a way that their activities could be matched. Indeed, such matching is a critical part of learning. If the cortex is concerned with the categorization of the world and the limbic–brain stem system is concerned with value (or with setting its adjustments to evolutionarily selected physiological patterns), then learning may be seen as the means by which categorization occurs on a background of value to result in adaptive changes in behavior that satisfy value.

Learning certainly occurs in animals that show no evidence of conscious behavior. But in some animal species with cortical systems, the categorizations of separate causally unconnected parts of the world can be correlated and bound into a *scene*. By a scene I mean a spatiotemporally ordered set of categorizations of familiar and nonfamiliar events, *some with and some without necessary physical or causal connections to others in the same scene*. The advantage provided by the ability to construct a scene is that events that may have had significance to an animal's past learning can be related to new events, however causally unconnected those events are in the outside world. Even more importantly, this relationship can be established in terms of the demands of the value systems of the individual animal. By these means, the salience of an event is determined not only by its position and energy in the physical world but also by the relative value it has been accorded in the past history of the individual animal as a result of learning.

It is the evolutionary development of the ability to create a scene that led to the emergence of primary consciousness. Obviously, for that emergence to have survived, it must have resulted in increased fitness. But before considering how, let's consider the model itself.

The appearance of primary consciousness, according to the model, depends on the evolution of three functions. Two of these evolutionary developments are necessary but not sufficient for consciousness. The first is the development of the cortical system in such a way that when conceptual functions appeared they could be linked strongly to the limbic system, extending already existing capacities to carry out learning. The second is the development of a new kind of memory based on this linkage. Unlike the system of perceptual categorization, this conceptual memory system is able to *categorize responses in the different brain systems* that carry out perceptual categorization and it does this according to the demands of limbic–brain stem value systems. This "value-category" memory allows conceptual responses to occur in terms of the *mutual* interactions of the thalamocortical and limbic–brain stem systems.

A third and critical evolutionary development provides a sufficient means for the appearance of primary consciousness. This is a special reentrant circuit that emerged during evolution as a new component of neuroanatomy. This circuit allows for continual reentrant signaling between the value-category memory and the ongoing global mappings that are concerned with perceptual categorization in real time. An animal without these new reentrant connections can carry out perceptual categorizations in various sensory modalities and can even develop a conceptual value-category memory. Such an animal cannot, however, link perceptual events into an ongoing scene. With the appearance of the new reentrant circuits in each modality, *a conceptual categorization of concurrent perceptions* can occur *before* these perceptual signals contribute lastingly to that memory. This interaction between a special kind of memory and perceptual categorization gives rise to primary consciousness. Given the appropriate reentrant circuits in the brain, this "bootstrapping process" takes place in all sensory modalities in parallel and simultaneously, thus allowing for the construction of a complex scene. The coherence of this scene is coordinated by the conceptual value-category memory even if the individual perceptual categorization events that contribute to it are causally independent.

My use of the word "scene" is meant to convey the idea that responses to roughly contemporaneous events in the world are connected by a set of reentrant processes. As human beings possessing higher-order consciousness, we experience primary consciousness as a "picture" or a "mental image" of ongoing categorized events. But as we shall see when we examine higher-order consciousness, there is no actual image or sketch in the brain. The "image" is a *correlation* between different kinds of categorizations.

To summarize: The brain carries out a process of conceptual "self-categorization." Self-categories are built by matching past perceptual categories with signals from value systems, a process carried out by cortical systems capable of conceptual functions. This value-category system then interacts via reentrant connections with brain areas carrying out ongoing perceptual categorizations of world events and signals. Perceptual (phenomenal) experience arises from the correlation by a conceptual memory of a set of ongoing perceptual categorizations. Primary consciousness is a kind of "remembered present."

These notions are illustrated in figure 11–1. While the diagram hardly conveys the complexity of the neural circuits involved, it does highlight several points. The first concerns what we may call self and nonself components. (By self in this context I mean a unique biological individual, not a socially constructed "human" self.) The self, or internal systems, arise from interactions between the limbic and the cortical systems. This differentiates them from outside-world systems that are strictly cortical.

The second point concerns the formation of value-category memory.

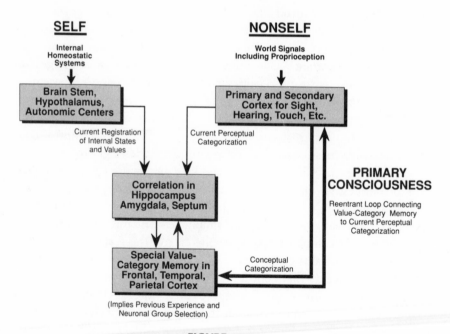

FIGURE 11–1

A model of primary consciousness. Past signals related to value (set by internal control systems) and categorized signals from the outside world are correlated and lead to memory in conceptual areas. This memory, which is capable of conceptual categorization, is linked by reentrant paths to current perceptual categorization of world signals (heavy lines). This results in primary consciousness. When it occurs through many modalities (sight, touch, and so forth), primary consciousness is of a "scene" made up of objects and events, some of which are not causally connected. An animal with primary consciousness can nonetheless connect these objects and events through memory via its previous value-laden experience.

This conceptual memory depends on constant interaction between self and world systems. The third point concerns the occurrence in real time and *in parallel* of perceptual categorizations for *each* sensory modality via the cortical system, including the organs of succession. The final and critical point heralds the appearance of primary consciousness: A correlative scene results from the function of reentrant connections between those cortical systems mediating conceptual value-category memory and those thalamo-cortical systems mediating ongoing perceptual categorizations across all the senses.

Notice that primary consciousness as I have characterized it has the necessary Jamesian properties: It is individual ("self" systems contribute to it), it is continuous and yet changing (as both world and internal signals evolve), and it is intentional (referring necessarily to internally given or outside-world signals derived alternately from things and events). If figure 11–1 were to be reiterated in a series of time steps, it would serve to stress these Jamesian properties of primary consciousness and the kind of perceptual bootstrapping that primary consciousness represents. Jamesian properties stress the flow of consciousness, its "before" and "after." In the conscious process, current value-free perceptual categorization interacts with value-dominated memory. This occurs *before* perceptual events contribute further to the alteration of that memory. When such events do contribute to the alteration of that memory, they are, in general, no longer in the specious or remembered present, that is, they are no longer in primary consciousness.

What is the evolutionary value of such a system? Obviously, primary consciousness must be efficacious if this biological account is correct. Consciousness is not merely an epiphenomenon. According to the TNGS, primary consciousness helps to abstract and organize complex changes in an environment involving multiple parallel signals. Even though some of these signals may have no direct causal connection to each other in the outside world, they may be significant indicators *to the animal* of danger or reward. This is because primary consciousness connects their features in terms of the saliency determined by the animal's past history and its values.

Primary consciousness provides a means of relating an individual's present input to its acts and past rewards. By presenting a correlative scene, it provides an adaptive way of directing attention during the sequencing of complex learning tasks. It also provides an efficient means of correcting errors. These performances might conceivably be carried out without the construction of a scene. But it seems likely that an animal with primary consciousness would have the ability to generalize its learning abilities across many more cues more quickly than an animal without it. Consciousness, I repeat, is efficacious and likely to enhance evolutionary fitness.

Primary consciousness is required for the evolution of higher-order consciousness. But it is limited to a small memorial interval around a time chunk I call the present. It lacks an explicit *notion* or a concept of a personal self, and it does not afford the ability to model the past or the future as part of a correlated scene. An animal with primary consciousness sees the room the way a beam of light illuminates it. Only that which is in the beam is explicitly in the remembered present; all else is darkness. This does not mean that an animal with primary consciousness cannot have long-term memory or act on it. Obviously, it can, but it cannot, in general, be aware of that memory or plan an extended future for itself based on that memory.

Where are the actual brain loci mediating primary consciousness? I have written elsewhere about the possibility that certain circuits in the thalamus, between the cortex and the thalamus, and connecting one cortical region to another may be the sites of the key reentrant circuits. I will not overload this discussion with the actual neuroanatomy (see figure 11–1 for the names of the areas involved). Nevertheless, it may be useful to mention here that, as revealed by cognitive testing, certain brain lesions lead to the selective loss of the explicit *conscious* recognition of a signal within a given perceptual domain that is nonetheless implicitly recognized, as shown by psychological tests of the affected person.

A good example is provided by stroke patients who have prosopagnosia—the inability to recognize faces as such. Although they have no awareness of faces, some of these patients will, while denying that they recognize their spouse's face, perform on tests in such a way as to indicate strong discriminatory knowledge of that face. Another example is blind sight. Individuals with lesions in their primary visual cortex report blindness—no awareness of vision—but can locate objects in space when tested. These matters will be discussed further in chapter 18. I mention them here to point out that they may be explained by assuming disruptions (within the appropriate perceptual domains) of the reentrant loops that I have postulated as important for primary consciousness (figure 11–1). Let us defer the discussion of tests for consciousness until later.

Before turning to the development of higher-order consciousness, a few words about some sticky matters are in order. The first is: Which animals have primary consciousness? I really cannot answer this except by relating it to the human referent that we agreed on. Going backward from the human referent, we may be reasonably sure (for reasons that will be made clear later) that chimpanzees have it. In all likelihood, most mammals and some birds may have it, although we can only test for its presence indirectly. Unfortunately, such tests are only neuroanatomical or behavioral (not by sign communication or report). If the brain systems required by the present model represent the *only* evolutionary path to primary consciousness, we can be fairly sure that animals without a cortex or its equivalent lack it. An amusing speculation is that cold-blooded animals with primitive cortices would face severe restrictions on primary consciousness because their value systems and value-category memory lack a stable enough biochemical milieu in which to make appropriate linkages to a system that could sustain such consciousness. So snakes are in (dubiously, depending on the temperature), but lobsters are out. If further study bears out this surmise, consciousness is about 300 million years old.

CHAPTER 12

Language and Higher-Order Consciousness

Human consciousness is a perpetual pursuit of a language and a style. To assume consciousness is at once to assume form. Even at levels below the zone of definition and clarity, measures and relationships exist. The chief characteristic of the mind is to be constantly describing itself.

—Henri Focillon

The last two chapters have involved a strenuous march through variegated and difficult terrain. But if you will bear with me through the next march, I believe you will be able to look back and see things more clearly—to make things "click." This is not quite possible at this juncture—to "see" clearly how primary consciousness works requires seeing how higher-order consciousness emerges and differs from it.

It is curious that we, as human beings with higher-order consciousness, cannot "see the world" with our primary consciousness alone. Creatures with primary consciousness, while possessing mental images, have no capacity to view those images from the vantage point of a socially constructed self. Yet one who has such a self as a result of higher-order consciousness *needs* it to link one mental image to the next in order to appreciate the workings of primary consciousness! Higher-order consciousness cannot be abandoned without losing the descriptive power it makes possible. (I often wonder whether this abandonment is what some mystics seek.)

124

What we can usefully do before taking up the origins of higher-order consciousness is to see what "function" there is in our proposed models for various kinds of categorization. Perceptual categorization, for example, is nonconscious and can be carried out by classification couples, or even by automata. It treats *signals from the outside world*—that is, signals from sensory sheets and organs. By contrast, conceptual categorization works from within the brain, requires perceptual categorization and memory, and treats *the activities of portions of global mappings* as its substrate. Connecting the two kinds of categorization with an additional reentrant path for each sensory modality (that is, in addition to the path that allows conceptual learning to take place) gives rise in primary consciousness to a correlated scene, or "image." This image can be regenerated in part by memory in animals with primary consciousness, but it cannot be regenerated in reference to a *symbolic* memory. By this I mean a memory for symbols and their associated meanings. And so an animal with primary consciousness alone is strongly tied to the succession of events in real time.

How can the tyranny of this remembered present be broken? The imprecise answer is: By the evolution of new forms of symbolic memory and new systems serving social communication and transmission. In its most developed form, this means the evolutionary acquisition of the capability for language. Inasmuch as human beings are the only species with language, it also means that higher-order consciousness has flowered in our species. But there are strong indications that we can see at least some of its origins in chimpanzees. Both species can think, not just have concepts, and chimpanzees also appear to have some elements of a self-concept. Certainly, the *basis* for recognizing a subject–predicate relationship in humans requires an emerging consciousness of the distinction between the self (in the social sense of "selfhood") and other entities classified as nonself. Chimpanzees have behaviors indicating that they make the distinction, but they lack true language and so I claim that what I call higher-order consciousness cannot flourish in them as it does in us.

Higher-order consciousness obviously requires the continued operation of the structures serving primary consciousness. In addition, it involves the ability to construct a socially based selfhood, to model the world in terms of the past and the future, and to be directly aware. Without a symbolic memory, these abilities cannot develop.

To trace how these abilities may have developed through the evolutionary emergence of a symbolic memory, it will be necessary to consider how speech evolved and how it is acquired. Therefore, I will first consider how the emergence of true language required the evolution of the vocal tract and the brain centers for speech production and comprehension. I then will confront an issue central to this essay: whether concepts are formed prior to speech. In doing so, I will conclude that a model of self—nonself interaction probably had to emerge prior to true speech.

SPEECH: AN EPIGENETIC THEORY

The considerations presented so far suggest that a model for speech acquisition requires primary consciousness. Furthermore, the development of a rich syntax and grammar is highly improbable without the prior evolution of a neural means for concepts. If this turns out to be true, it will be obvious why computers are unable to deal with semantic situations. Their embodiment is wrong; it does not lead to consciousness.

I propose that before language evolved, the brain already had the necessary bases for meanings in its capacities to produce and act on concepts. The evolution in primates of rich conceptual memories, and in hominids of phonological capabilities and special brain regions for the production, ordering, and memory of speech sounds, then opened up the possibility of the emergence of higher-order consciousness. (Although I will not discuss the details of grammatical systems here, a pertinent discussion of some aspects of grammar may be found in the Postscript.)

Speech is special and unique to *Homo sapiens*. Can we account for its evolutionary emergence without creating a gulf between linguistic theory and biology? Yes, provided that we account for speech in epigenetic as well as genetic terms. This means abandoning any notion of a genetically programmed language-acquisition device. It does not mean, however, that specialized heritable structures were not necessary for speech to arise. Indeed, the evidence for the existence of specialized heritable structures related to speech is not hard to find. After the assumption of bipedal posture by hominids, changes occurred in the basicranial structure of their skulls (figure 12–1). This provided a morphological basis for the evolution of a uniquely human piece of anatomy, the supralaryngeal tract or space. This tract becomes mature in human infants when the larynx descends. (To avoid choking, a structure called the epiglottis must close when humans eat. Indeed, unlike other animals, we cannot phonate and swallow at the same time without potential disaster.) As part of this evolutionary development, the vocal folds emerged and the tongue, palate, and teeth were selected to allow fuller control of air flow over the vocal cords, which in turn allowed the production of coarticulated sounds, the phonemes.

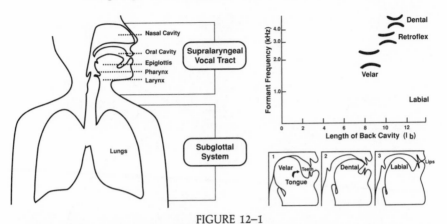

FIGURE 12–1

The supralaryngeal tract in humans, one of the main anatomical bases for speech production and the evolution of language. This complex structure functions because the larynx (voice box) has descended so that exhaled air effectively causes the vibration of the vocal cords, which are in turn altered in tension and apposition by exquisite muscular changes (left). Modulation by other elements—the tongue, teeth, lips, and so forth—leads to a set of coarticulated sounds (right top). There is a risk of choking to death if the epiglottis doesn't close the airway when swallowing occurs (right bottom). These changes presuppose that changes have already occurred in the base of the skull during evolution.

At the same time or shortly after in evolution, special cerebral cortical regions emerged on the left side that are now known as Broca's and Wernicke's areas (figure 12–2). These cortical regions linked acoustic, motor, and conceptual areas of the brain by reentrant connections. Through these connections, Broca's and Wernicke's areas served to coordinate the production and categorization of speech. Most importantly, they provided a system for the development of a new kind of memory capable of recategorizing phonemes (the basic units of speech) as well as their order.

We may reasonably assume that phonology arose in a speech community that used primitive sentences (perhaps resembling present-day pidgin languages) as major units of exchange. In such an early community, utterances correlated nouns with objects and led to the beginnings of semantics (figure 12–3). Verbs followed. Note that the preexisting capacity for concepts provided a necessary basis for these semantic developments. In early humans, the presyntactical organization of gestures may have allowed a simple ordering of nouns and verbs. Further development of Broca's and Wernicke's areas allowed the more sophisticated sensorimotor ordering that is the basis of true syntax.

According to the theory of speech acquisition that I favor, syntax

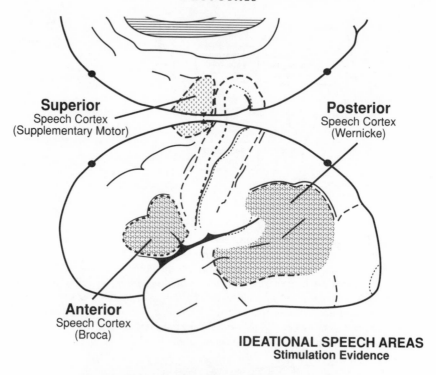

Superior
Speech Cortex
(Supplementary Motor)

Posterior
Speech Cortex
(Wernicke)

Anterior
Speech Cortex
(Broca)

IDEATIONAL SPEECH AREAS
Stimulation Evidence

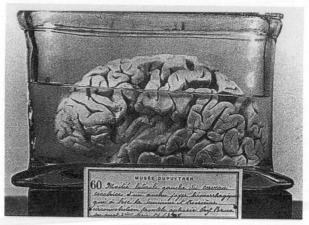

FIGURE 12–2

Areas of the brain serving speech production (top). If these brain regions are damaged, aphasia occurs in a variety of forms. Pictured is the brain of one of Paul Broca's patients who had a lesion in what is called Broca's area (bottom). Its owner, when alive, had motor aphasia.

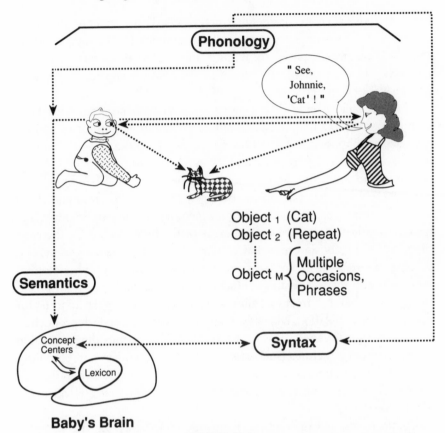

Baby's Brain

FIGURE 12–3

Semantic bootstrapping. A scheme crudely showing how affect, reward, and learning under conditions of categorization lead to speech acquisition. Phonology provides the means to connect categorized objects to semantics. As reentrant connections are made with concept centers, semantic bootstrapping occurs. As a lexicon is built and sentences are experienced, the categorization of their arrangements leads to syntax.

emerged epigenetically in a definite order (figure 12–3). First, phonological capabilities were linked by learning with concepts and gestures, which allowed for the development of semantics. This development permitted the accumulation of a lexicon: words and phrases with meaning. Syntax then emerged by connecting preexisting conceptual learning to lexical learning. A similar idea has been proposed by Steven Pinker and others within the framework of a grammar developed by Joan Bresnan, which she calls lexical functional grammar. They call this process semantic bootstrapping. In the extended TNGS, I provide explicit evolutionary, anatomical, and physiological arguments to support the notion that an infant already has conceptual categories that are *not* defined or originated by semantic means or criteria. These categories are required if semantic bootstrapping is to occur, and they are also required to support the related proposals of Ronald Langacker, George Lakoff, and others of what have been called "cognitive grammars" (see the Postscript).

Thus, to build syntax or the bases for grammar, the brain must have reentrant structures that allow semantics to emerge *first* (prior to syntax) by relating phonological symbols to concepts. Because of the special memory provided by Broca's and Wernicke's areas, the phonological, semantic, and syntactical levels can interact directly and also indirectly via reentrant circuits that are formed between these speech areas and those brain areas that subserve value-category memory. When a sufficiently large lexicon is collected, the conceptual areas of the brain categorize the *order* of speech elements, an order that is then stabilized in memory as syntax. In other words, the brain recursively relates semantic to phonological sequences and then generates syntactic correspondences, not from preexisting rules, but by treating rules *developing in memory* as objects for conceptual manipulation. Memory, comprehension, and speech production interact in a great variety of ways by reentry. This permits the production of higher-order structures (such as sentences in a grammar) and obviously helps with the elaboration of lower-order sequences (such as phrases). Of course, once achieved, the sequencing becomes automatic, as do many other motor acts.

Chimpanzees, unlike humans, have no brain bases for the complex sequencing of articulated sounds. They appear to have concepts and thought and are even capable of a simple "semantics," but inasmuch as they lack an elaborated syntax, they have no true language or speech per se.

It is obvious why the acquisition of true speech leads to an enormous increase in conceptual power. The addition of a special symbolic memory connected to preexisting conceptual centers results in the ability to elaborate, refine, connect, create, and remember great numbers of new concepts. It is not the case that the language centers "contain" concepts or that concepts "arise" from speech. Meaning arises from the interaction of value-category memory with the *combined* activity of conceptual areas and speech areas. And although vocal speech was probably necessary for the evolutionary selection of the necessary morphological changes in the brain, after their emergence any gestural system in a speech community (such as sign language) could be employed if necessary. Moreover, like many brain systems subject to epigenetic development, the system underlying speech acquisition differs in children and in adults; it is subject to a developmental critical period. In all likelihood, this time period is related to extensive synaptic and neuronal group selection occurring up to adolescence, after which time such changes occur much less extensively and in a different fashion.

This theory of speech is a nativist theory insofar as it requires the prior evolution of special brain structures. But it invokes no new principles beyond those of the TNGS. It is not a computational theory, nor one that insists on a language acquisition device containing innate genetically specified rules for a universal grammar. Syntax is built epigenetically under genetic constraints, just as human faces (which are about as universal as grammar) are similarly built by different developmental constraints. The principles of topobiology (see chapter 6) apply to both cases.

This proposal is compatible with the capacity to construct and interpret a potentially infinite number of sentences from a finite number of words. This is so because the generalizing and categorizing power of a conceptual system interacting reentrantly and recursively with specialized language areas is well-nigh unlimited. Inasmuch as syntax is constructed from semantics (under constraints), local grammatical relations can be constructed even from sentence fragments free of the strict order of sentences. A grammar so built is necessarily mapped onto the continual activities of a very definite set of brain structures, among which the most important may be those giving rise to primary consciousness. Indeed, if this theory is correct, language is impossible without primary consciousness.

HIGHER-ORDER CONSCIOUSNESS

With this theory of speech in hand, we may return to our main subject: higher-order consciousness. How does one become "conscious of being conscious?" In order to acquire this capacity, systems of memory must be related to a conceptual representation of a true self (or social self) acting on an environment and vice versa. A conceptual model of selfhood must be built, as well as a model of the past. A number of steps of developmental learning that alter the individual's relation to the immediate present are necessary for this to take place.

Brain repertoires are required that are able to delay responses. (Repertoires of this type are known to be present in the frontal cortex.) These repertoires must be able to categorize the processes of primary consciousness itself. This is achieved largely through symbolic means, by comparison and reward during social transmission and learning. During the acquisition of semantics, that reward arises by relating speech symbols to the gratification of affective needs by conspecifics in parental, grooming, or sexual interactions.

The figure (figure 12–4) showing the relation of speech areas to conceptual areas, which allows for the development of a concept of self and of higher-order consciousness, must be supplemented with one showing social relations (see figure 12–3). Long-term storage of symbolic relations, acquired through interactions with other individuals of the same species, is critical to the self-concept. This acquisition is accompanied by the categorization of sentences related to self and nonself and their connection to events in primary consciousness. The corresponding elaboration achieved by the learning of elements in phonemic and symbolic memories also allows more effective categorizations through verbs of various *acts* in relation to the self and others.

The interaction between this specialized set of memories and conceptual value-category memory allows for a modeling of the world. And given the

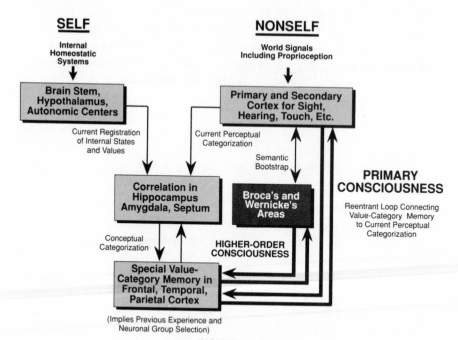

FIGURE 12–4

A scheme for higher-order consciousness. (The reader may relate this to the scheme shown in figure 11–1 for primary consciousness.) The acquisition of a new kind of memory via semantic bootstrapping (figure 12–3) leads to a conceptual explosion. As a result, concepts of the self, the past, and the future can be connected to primary consciousness. "Consciousness of consciousness" becomes possible.

emergence of the ability to distinguish such conceptual–symbolic models from ongoing perceptual experience, a concept of the past can be developed. This frees the individual from the bondage of an immediate time frame or ongoing events occurring in real time. The remembered present is placed within a framework of past and future.

While the embodiment of meaning and reference can be related to real objects and events by the reentrant connections between value-category memory and perception (primary consciousness), simultaneous interactions can also occur between a symbolic memory and the same conceptual centers. An inner life, based on the emergence of language in a speech community, becomes possible. This is tied to perceptual and conceptual structures, but it is highly individual (indeed, it is personal) and it is also strongly tied to affect and reward. It is higher-order consciousness, capable of modeling the past, present, future, a self, and a world.

One of the astonishing features of higher-order consciousness is how rapidly it appeared. Paleontological studies have shown that these developments occurred over very short periods of evolutionary time. The topobiological principles underlying brain development and the mechanisms of the TNGS can account for this rapid emergence, for they allow for enormous changes in brain size over the relatively short evolutionary periods in which *Homo sapiens* emerged. According to topobiology, morphological changes of relatively large extent occur through changes in the timing of the action of morphoregulatory genes as a result of relatively few mutations (see chapter 6). And the premises of the TNGS allow for the rapid incorporation of new and enlarged primary repertoires into existing brain structures.

A synoptic picture of how consciousness is related to evolutionary morphology is diagrammed in figure 12–5. While this hardly gives the details and even lacks a time frame, it does suggest how *two successive sets of bootstrapping events* (perceptual and semantic), each involving the evolution of new morphology (memory circuits and new forms of reentry) could give rise first to primary consciousness and then to higher-order consciousness.

This evolutionary panorama provokes additional questions about the adaptive advantages of consciousness. Primary consciousness provides the ability to determine by internal criteria the salience of patterns among multiple parallel signals arising in complex environments. That salience is largely but not completely determined by the previous history and learning of the individual animal. Higher-order consciousness adds socially constructed selfhood to this picture of biological individuality. The freeing of parts of conscious thought from the constraints of an immediate present and the increased richness of social communication allow for the anticipa-

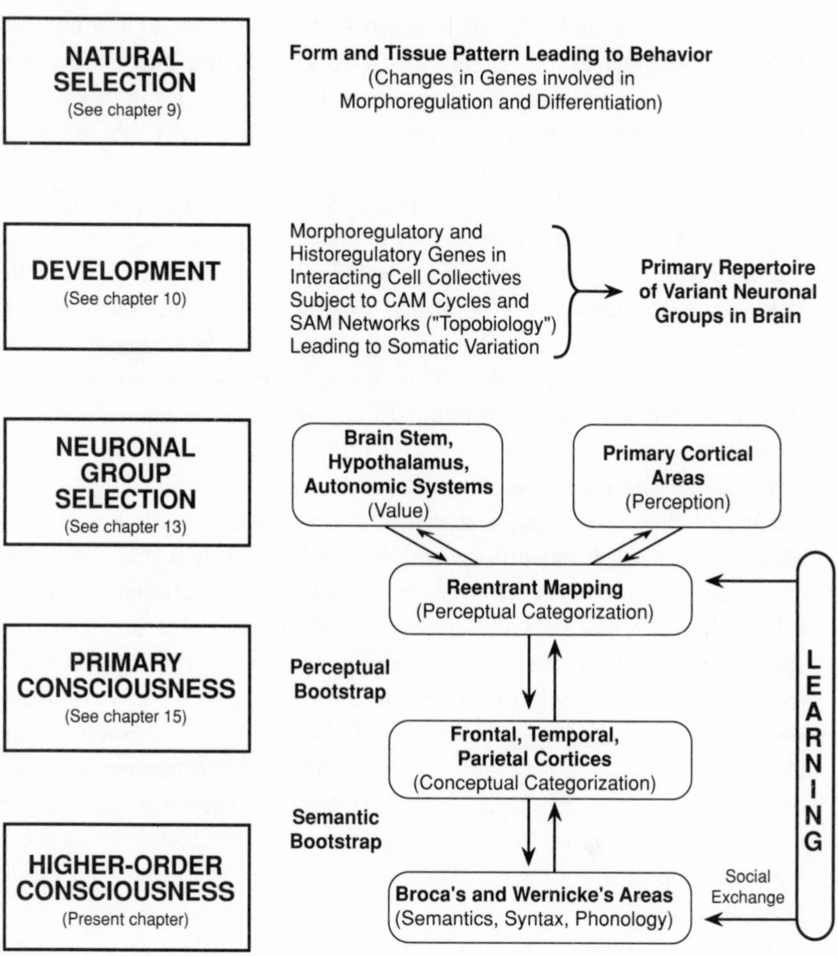

NATURAL SELECTION (See chapter 9)	**Form and Tissue Pattern Leading to Behavior** (Changes in Genes involved in Morphoregulation and Differentiation)

DEVELOPMENT (See chapter 10)	Morphoregulatory and Historegulatory Genes in Interacting Cell Collectives Subject to CAM Cycles and SAM Networks ("Topobiology") Leading to Somatic Variation	**Primary Repertoire of Variant Neuronal Groups in Brain**

NEURONAL GROUP SELECTION (See chapter 13)

Brain Stem, Hypothalamus, Autonomic Systems (Value)

Primary Cortical Areas (Perception)

Reentrant Mapping (Perceptual Categorization)

PRIMARY CONSCIOUSNESS (See chapter 15)

Perceptual Bootstrap

Frontal, Temporal, Parietal Cortices (Conceptual Categorization)

Semantic Bootstrap

HIGHER-ORDER CONSCIOUSNESS (Present chapter)

Broca's and Wernicke's Areas (Semantics, Syntax, Phonology)

Social Exchange

LEARNING

FIGURE 12–5

The evolution of consciousness depends on the evolution of new morphology. Here, an evolutionary sequence of events is shown in which the principles of natural selection and development lead to neural recognition systems and result in conscious experience. No new principles besides those of the theory of neuronal group selection are required. But newly evolved anatomical structures selected for function are required. These include those shown in the first two figures of this chapter. The principles underlying the function of each area are explained in the chapters indicated. Notice that a "perceptual bootstrap" produces primary consciousness and a "semantic bootstrap" produces higher-order consciousness. Both bootstraps rely on the evolution of appropriate reentrant pathways in the brain.

tion of future states and for planned behavior. With that ability come the abilities to model the world, to make explicit comparisons and to weigh outcomes; through such comparisons comes the possibility of reorganizing plans. Obviously, these capabilities have adaptive value. The history of humanity since the evolution of hunter-gatherers speaks to both the adaptive and the maladaptive properties of the only species with fully developed higher-order consciousness.

Some anthropologists have proposed (somewhat fancifully) that our brains enlarged so rapidly because, after a certain point, higher-order consciousness conferred on us the ability to deceive ourselves in such a way as to allow us to deceive others more "sincerely" to our own advantage. In a socially bound animal, this deceptiveness might, according to these authors, have selective advantages. An old tale tells of Boris and Ivan at the train station. Ivan says, "Boris, where are you going?" Boris replies, "To Minsk." Ivan then says, "Boris, I know you. If you were going to Minsk, you would tell me you were going to Pinsk. Now, I happen to know you are going to Minsk. So, why are you lying to me?"

Given this picture of a human animal simultaneously and interactively capable of higher-order and primary consciousness, we may return to the vexing question of qualia. Recall that our theoretical analysis of consciousness was founded on three assumptions: the physics assumption, the evolutionary assumption, and the qualia assumption. Having already assumed that human beings have qualia, why return to the issue? We know that a God's-eye view—in which the theory would, through the communication of its structure, allow an imaginary qualia-free animal know what qualia are—is not feasible. We have said enough about the mechanisms of consciousness to indicate that only through direct possession by an individual of the appropriate morphology and experience do qualia arise. Nevertheless, our elaborated picture provides certain refinements.

First of all, it is clear how *different* qualia are *discriminated*—through differences in neural structure and behavior in different sensory pathways. This has been known for a long time, since the doctrine of specific nerve energies advanced by Johannes Müller. We can add that an animal with higher-order consciousness is likely to call a given phenomenal state by a different name than another one evoked by a different neural pathway (if not "green" or "warm," then at least "ween" and "grarm," but in any case, in a consistent fashion).

If animals having only primary consciousness also have qualia, they cannot report them explicitly either to a human observer or to themselves, for they lack conceptual selves. Like flashlights illuminating a room, their qualia, if they occur, exist only for the duration of the remembered present of the scene. We can only adduce their possible presence by observing the behavioral responses of these animals.

But with us, it is different. Qualia, individual to each of us, are recategorizations by higher-order consciousness of value-laden perceptual relations in each sensory modality or their conceptual combinations with each other. We report them crudely to others; they are more directly reportable to ourselves. This set of relationships is usually but not always connected to value. *Freedom from time allows the location in time* of phenomenal states by a suffering or joyous self. And the presence of appropriate language improves discrimination enormously; skill in wine tasting, for example, may be considered the result of a passion based on qualia that are increasingly refined by language.

With this view of higher-order consciousness, it is possible to see roughly what lies beneath the self that connects phonology to semantics in a naming sentence. Once a self is developed through social and linguistic interactions on a base of primary consciousness, a world is developed that *requires* naming and intending. This world reflects inner events that are recalled, and imagined events, as well as outside events that are perceptually experienced. Tragedy becomes possible—the loss of the self by death or mental disorder, the remembrance of unassuageable pain. By the same token, a high drama of creation and endless imagination emerges.

Ironically, the self is the last thing to be understood by its possessor, even after the possession of a theory of consciousness. Given the way in which higher-order consciousness arises and naming occurs, that should be no surprise, except to each of us as a possessor. Embodiment imposes ineluctable limits. The wish to go beyond these limits creates contradiction, fantasy, and a mystique that makes the study of the mind especially challenging, for after a certain point, in its individual creations at least, the mind lies beyond scientific reach. Scientific study recognizes this limit without indulging in mystical exercises or illusions. The reason for the limit is straightforward: The forms of embodiment that lead to consciousness are unique in each individual, unique to his or her body and individual history.

CHAPTER 13

Attention and the Unconscious

Illusions commend themselves to us because they save us pain and allow us to enjoy pleasure instead. We must therefore accept it without complaint when they sometimes collide with a bit of reality against which they are dashed to pieces.

—Sigmund Freud

Consciousness reigns but doesn't govern.

—Paul Valéry

There are two objections that may be raised to what I have said so far. The first is that I have not *really* explained "what it is like" to be conscious. The second is that there appears to be much I have failed to explain, for example, the fact that a good deal of our behavior is unconsciously driven. Sigmund Freud spent much of his life trying to understand this, particularly trying to understand the repression of experiences that were threatening, painful, or unpleasant. I will consider the first objection briefly and the second at greater length.

Given the central importance of consciousness to knowing that we exist—Descartes' claim—it is no surprise that much would be expected of any account that presumes to explain it. Of the present one it might be said, "You may think you have explained how memory, perceptual categorization, reentry, and so on, work to give the *properties* of consciousness, but you have not explained how *I* feel being conscious, or why *I* feel myself to be conscious. Consciousness is strange, mysterious, the ultimate mystery." To reply I have to point out the limits of any claim to scientific explanation and then show what is special about any explanation proposed for consciousness.

Science is concerned with the formal correlations of properties, and with the development of theoretical constructs that most parsimoniously and usefully describe all known aspects of that correlation, without exception. It must couch its descriptions in terms that can be exchanged and understood between any two human observers. Any description that does not assume a conscious, understanding human observer as its target, one who can object to flaws in logic, repeat experiments, and construct new ones, is not a scientific description. An example of a nonscientific description is a personal account of my particular sensations, apparent recollections, and emotions during a drug-induced trance. At best, someone may correlate the reports of twenty subjects (including me) in such a trance and find regularities. But he or she will not be able to correlate reliably *my* actual feelings, particular history, and mode of forgetting in any detail or with any general certainty. So science fails for individual histories even though it may succeed in discerning what is common among twenty chronicles.

There *is* something peculiar about consciousness as a subject of science, for consciousness itself is the individual, personal process each of us must possess in working order to proceed with *any* scientific explanation. Even though I may be unaware of what I have forgotten or repressed, or of unconscious factors that drive my behavior, I feel as if the process of consciousness is all of a piece, at least in my healthy state. And so it is natural that I demand an explanation of my own consciousness in terms satisfactory to myself. But I must realize that it is not a scientific act to do so, nor would I expect it to be. After all, no one says to a physicist, "You have explained energy and matter in terms of symmetry relations, and you have even approached the beginnings of the universe in your theories. But you have not *really* explained *why* there is something rather than nothing." To attempt such an explanation would be fruitless; under these circumstances, no science based on experiment could recommend itself as better than any other. A scientific explanation cannot be given.

Well then, why are we tempted to demand a scientific explanation of how it feels personally to be conscious? It is the certainty of consciousness to ourselves and its relation to the idea of self that makes us want to demand more of a psychologist than of a physicist or a cosmologist. But the demand is not a scientifically reasonable one.

A reply to the question would have exactly the same form as one given by a physicist would have. "I have offered you a theory in terms of known structures and relationships, one based on experimental facts. The theory says that, if you perform an operation on structures said to be important for properties of consciousness, those properties will be predictably altered or may even disappear. If, for example, you cut a reentrant loop connecting a part of the brain essential for carrying out face recognition (and just that part), a conscious person will experience prosopagnosia. That person will become and remain *unconscious* of the fact that he or she can (as shown by examination of implicit memory) still recognize faces that an observer knows the person has seen before, and he or she will quite sincerely deny recognizing them. So it is with other tests; a theory of consciousness *can* have operational components. But it must, if it is a good theory, also unify all kinds of pertinent facts and deepen our understanding. (For example, it should explain how some processes can be unconscious and yet motivate behavior, which is the task of this chapter.)

Why then do we insist on a God's-eye view, even in the face of these explanations? Why is there a consciousness mystique—a desire for universal explanation, for conservation of consciousness as an individual experience, time without end? A reasonable answer seems to be that each consciousness depends on its unique history and embodiment. And given that a human conscious self is constructed, somewhat paradoxically, by social interactions, yet has been selected for during evolution to realize mainly the aims and satisfactions of each biological individual, it is perhaps no surprise that as individuals we want an explanation that science cannot give. It is also perhaps no surprise that we desire immortality. But there is no more mystery to our inability as scientists to give an explanation of an individual consciousness than there is to our inability to explain why there is something rather than nothing. There is a mystery perhaps, but it is not a scientific one. If one stays solely with one's own mind, the mystery rests in imagining how that particular mind arises with regard to its own personal history. We are "locked in."

There is one real but remote possibility of dealing scientifically with the "locked-in" property of consciousness in an individual, the source of this "mystery." If an artifact could be built that had structures and experiences allowing it both to become conscious and to have language, one could test for the presence and absence of qualia. If it reported a certain feeling, would it be reasonable (and ethical) to rebuild it without the structures surmised to be essential to that feeling? After such a procedure would the same artifact then feel "strange" and "different," given that its self would have been built from the interaction of unconscious processes and conscious "social" interactions? We must wait and see, but as fantastic and as improbable as such a proposal sounds, it is at least theoretically possible. The same cannot be said of experimenting with the creation of the universe.

Before turning to the unconscious processes that give rise to and alter consciousness, it may be useful to consider what is difficult to imagine about the mind and what is easy. As we have seen, one overriding difficulty haunts any attempt to explain the mind: It is that the mind arises as a result of physical interactions across an enormously large number of different levels of organization, ranging from the molecular to the social. Furthermore, these interactions are often idiosyncratic or irreversible, and the structural features central to their workings include parallel, one–many, or many–many mappings. Our brains (and particularly philosophers' brains) are not very good at visualizing such complex orderings. But the situation may not be hopeless; as I discuss later, the advent of increasingly powerful computers may help us build heuristics that can let us see how things go together.

Until this field is more fully developed, we may ask: What is easy to imagine about the mind? I believe most people would agree to the following list:

1. The workings of brain circuits *grosso modo* in terms of their inputs and outputs. The example is classical neurophysiology.
2. The interaction between an animal's pattern of behavior and the physical world of stimuli. The example is descriptive psychology.
3. Certain acts of social transmission. The example is the study of imprinting in ethology and the accepted propositions of folk psychology—what people seem to believe, desire, or intend.

What is hard to understand?

1. The net result of the simultaneous action in parallel of complex neural populations. An example is the difficulty of predicting the outcome of the activity of a large number of neuronal groups.
2. Memory as a dynamic process and system property, one that is *not* equivalent to the sum of the synaptic changes that underlie it. An example is the overall response of an automaton like Darwin III after training.
3. More complex psychological phenomena such as consciousness. Numerous examples have been given in the last two chapters.
4. The idea of a socially constructed self resulting from the interactions of both unconscious and conscious processes. The example is discussed in the rest of this chapter.

Undoubtedly, one may be able to think of alternative lists. To understand psychological processes (particularly those in the second list) in terms of a brain theory, however, one must not only have a good theory but also be able to use synthetic computer models to check the self-consistency of its mechanisms and analyze multilevel interactions. Unfortunately, mathematics alone will not sufficiently aid our verbal constructions as it does our physics. At present, computer simulations seem more promising.

In considering this chapter, I suggest that the reader make every effort to understand how classification couples together with reentry yield perceptual categorization and how a dynamic memory functions as a system property (see chapters 9 and 10). With these two processes understood, I believe the reader can move step by step through the models I have described for primary and higher-order consciousness and "see" how they might yield Jamesian properties, change salience in a scene, and allow plans to be formulated. Perhaps the best way to try this is by examining the figures in the appropriate chapters. The understanding gained will make it much easier to see how attention and the unconscious operate.

ATTENTION

My account of consciousness has not explicitly dealt with attention, which James called "the taking possession by the mind, in clear and vivid form, of one out of what seem several simultaneously possible objects or trains of thought." Attention is not the same as consciousness, but its relationship to consciousness poses some of the most difficult problems for theory. Attention must, for example, be discriminated from overall wakefulness, for it is not simply a matter of vigilance or alertness; it lends a *directional* component to behavior, and it modulates an animal's responsiveness to the environment. Indeed, attention reveals the "fragility" of consciousness: It focuses our mind on its objects and obliterates or attenuates surrounding "irrelevancies." It does not seem possible to pay specific attention to more than a few objects or lines of thought; attention is highly selective, apparently obligatorily so.

Many theories of selective attention are based on the notion of "filtering out" input signals, either early or late. But there is a variety of evidence suggesting that such filtering does not occur. I have favored the notion, posited by others, that brain mechanisms of attention were originally derived from evolutionary pressure on an animal to select one out of a set of appropriate actions. An animal that is hungry or being threatened has to select an object or an action from many possible ones. It is obvious that the ability to choose quickly one action pattern to be carried out to the exclusion of others confers considerable selective advantage. Possessing such an ability makes it possible to achieve a goal that would otherwise be interfered with by the attempt to undertake two incompatible actions simultaneously. Survival may depend critically on this ability.

This "motor" theory of the origin of attention does not imply that perceptual components are not important. Indeed, it is obvious that the mechanisms of attention are multiple, ranging from perceptual competition to volitional choice. But if the end result is the formulation of a sequence of actions or motor plans, whether executed or not, then, according to the TNGS, global mappings and the basal ganglia are likely to be involved (see chapters 9 and 10, particularly figure 10–2). In an animal with primary consciousness, a balance must be struck between responding to internally determined salience and externally produced novelty. With higher-order consciousness the situation becomes much richer. Volitional states related to the selection of plans, values, and temporal projections can all change the relative contribution of different parts of a global mapping. In both cases, large portions of the nervous system are likely to be involved when a global mapping is modified to alter attention.

According to this view, we would expect attention to be altered by changes in several levels of a global mapping: by unconscious as well as by conscious activity. How, specifically, could such a system work? Any model proposed to explain attention must account for its selectivity; for the fact that, after an animal learns a skill, it becomes automatic; for the ability to interrupt automatic acts by attention to novelty; and for the ability to direct attention specifically by conscious means.

We all know that conscious attention plays a large role in the learning of complex skills. But in many cases, successful learning allows us to carry out skilled actions without attention. Performance then remains unconscious until either novelty or threat makes additional demands. Remember that, in presenting the TNGS, I suggested that the basal ganglia were major organs of succession, acting with the cortex to choose motor plans. Motor plans, which may be consciously formed in humans, are executed via the motor cortex as it sends signals to the spinal cord. But the output of the cortex is also routed to the basal ganglia. These structures have only an indirect connection back to the cortex, but it is a very significant one. The output from the basal ganglia is inhibitory, and therefore it can also *inhibit inhibition.* In other words, it can *disinhibit* target areas in the cortex. This either excites them or prepares them for excitatory input, a state important for attention.

In accord with a given plan, the basal ganglia selectively disinhibit thalamic nuclei projecting to the cortex. This leads to anticipatory and selective arousal of cortical areas corresponding to the motor program. These cortical areas then become more sensitive to those sensory inputs that are consistent with the performance of the task via a global mapping. Such a mechanism can explain focused attention.

What about automatic activity interrupted by novelty? If the task is not completed within a certain time, or if a novel event is detected and categorized, "alarm" signals may pass down to the midbrain value systems that connect back to the cortex and the basal ganglia. These systems may then send back signals to interrupt the motor plan in the cortex and block the execution of a motor program. As long as an automatized action is accomplished without a hitch, these midbrain nuclei are not engaged. Otherwise, as in the case of a shouted "watch out!" during a conversation while driving, they will cause a shift in attention to occur.

But how can *consciousness* alter attention and alter priorities in the construction of global mappings? In the case of primary consciousness, it can alter attention and priorities by a change in the salience among the parallel reentrant loops connected to the basal ganglia in a process similar to the one outlined above. In the case of higher-order consciousness, verbal schemas in conceptual areas can, through the activities of the frontal cortex and limbic system, dominate the apportionment of disinhibition by the basal ganglia, which have strong connections to such regions.

The fragility of attention is a particularly interesting issue. How is it that conscious attention is so narrow—usually able only to focus on one or at most two targets at the same time? An answer is suggested by the motor theory, which looks at attention as arising from evolutionary needs. Motor plans and programs are more or less exclusive (that is, they will not accommodate contradictory actions that are simultaneous). Moreover, given the large amount of nervous tissue involved in each global mapping, it seems unlikely that one could sustain more than a few complex mappings at the same time without their interfering with each other.

Such a view of attention still concedes a major overriding significance to nonconscious mechanisms and to the orienting behaviors mediated by global mappings in response to emergencies. Yet because having intentional conscious states depends on values, categories, and memories as well as plans, this selectional view of attention allows us the ability to entertain consciously an "intention to attend" to what is planned or envisioned. But this capacity is always subject to competition from unconscious and nonconscious elements (the latter being those that can *never* become conscious). We are all aware of parapraxes—slips of the tongue—and of actions committed "not as planned." These suggest the intervention of unconscious processes.

THE UNCONSCIOUS

Freud (figure 13–1) was the single most important figure in pointing up the role of unconscious processes in our behavior and feelings. In his "Project

FIGURE 13–1

Sigmund Freud (1856–1939), founder of psychoanalysis and explorer of the mechanisms of repression in memory.

for a Scientific Psychology," he tried to write an explicit neural account of the relation between conscious and unconscious processes and behavior, but soon abandoned the attempt. His later formulations were psychological explanations of behavior that emphasized intentionality but were at the same time ruthlessly deterministic.

The postulation of an unconscious is a central binding principle of Freud's psychological theories. Since his time, ample evidence has accumulated from the study of neurosis, hypnotism, and parapraxes to show that his basic theses about the action of the unconscious were essentially correct. As he used it, the term unconscious referred to elements that can be easily transformed into conscious states—"the preconscious"—as well as those that can be transformed only with great difficulty or not at all—"the unconscious proper." Freud suggested that threatening events could be repressed in memory so that they were not ordinarily available for conscious recall.

We must not forget that these are psychological, not structural, terms. My late friend, the molecular biologist Jacques Monod, used to argue vehemently with me about Freud, insisting that he was unscientific and quite possibly a charlatan. I took the side that, while perhaps not a scientist in our sense, Freud was a great intellectual pioneer, particularly in his views on the unconscious and its role in behavior. Monod, of stern Huguenot stock, replied, "I am entirely aware of my motives and entirely responsible for my actions. They are all conscious." In exasperation I once said, "Jacques, let's put it this way. Everything Freud said applies to me and none of it to you." He replied, "Exactly, my dear fellow."

Freud's notion of repression is consistent with the models of consciousness presented here. The extended TNGS strongly implicates value-dependent systems in memory formation. Self–nonself discrimination (see figure 11–1) requires the participation of memory systems that are forever inaccessible to consciousness. Repression, the selective inability to recall, would be subject to recategorizations that are strongly value-laden. And given the socially constructed nature of higher-order consciousness, it would be evolutionarily advantageous to have mechanisms to repress those recategorizations that threaten the efficacy of self-concepts. Circuitry that interacts with value systems exists in the hippocampus and the basal ganglia. In a linguistic animal, symbols matter, and the evolution of a way of reducing access to states considered threatening to the self-concept would have selective value. This provides a great clue to the properties of emotions, a subject to be touched on in a later chapter.

My general conclusion, important for all theories of mind, is that given the existence of acts driven by the unconscious, conclusions reached by conscious introspection may be subject to grave error. In other words, Cartesian incorrigibility is incompatible with the facts. Descartes, an adult genius with mastery over language, did not take several things into account. The first is the developmentally determined nature of higher-order consciousness. (Recall that French babies, even gifted ones, are unlikely to assert, "Je pense, donc je suis.") The second is that his linguistically based consciousness is *not* self-sufficient and beyond doubt. Given that it is linguistic, it is always in dialogue with some "other," even if that interlocutor is not present. The third is that unconscious mechanisms block and intervene with what we consider to be transparent and obvious lines of thought. By opening up the questions that prompted his method of doubt and by fearlessly exposing his thoughts on the nature of the mind, Descartes became a great pioneer of modern philosophical and psychological investigation. Since the announcement of his method, however, accumulated knowledge has forced us into a much humbler posture about the certainty of what we know.

This is probably a good point at which to state again how little it is we actually know. Given the difficulty of imagining the mind and the complex layerings of its workings and processes, we should not be surprised at this. But we have many resources open to us now that Descartes did not have. They should, in the end, allow us to appreciate how fruitful his "wrong" theory has been in stimulating our attempts to understand the mind.

CHAPTER 14

Layers and Loops: A Summary

It seems to me that the human race stands on the brink of a major breakthrough. We have advanced to the point where we can put our hand on the hem of the curtain that separates us from an understanding of the nature of our minds. Is it conceivable that we will withdraw our hand and turn back through discouragement and lack of vision?

—Percy Williams Bridgman

It is high time for another view of the mental, for a neuroscientific model of the mind. What makes the one proposed here new is that it is based remorselessly on physics and biology. It is also based on the ideas of evolutionary morphology and selection, and it rejects the notion that a syntactical description of mental operations and representations (see the Postscript) suffices to explain the mind. Others have held similar positions but have not united them in a single evolutionarily based theory, one that connects embryology, morphology, physiology, and psychology. Only such a physically based theory of mind is open to disconfirmation by scientific means.

The road connecting these disciplines is a bumpy one and, as the reader has seen, traveling the route is occasionally strenuous. This is because, in the construction of the mind, so many levels of organization are required and so many interactive loops have to be made to link what at first appear to be disparate layers of description. Given that the mind is a result of evolution and not of logical planning, I would not expect a different outcome. This profusion of levels, and not some esoteric new principle, whether of physics or of theosophy, is why it is difficult to think about the mind. The brain giving rise to the mind is a prototypical complex system, one more akin in its style of construction to a jungle than to a computer. This analogy fails at one point: While plants in jungles are selected for during evolution, the jungle itself is not. But the brain is subjected to two processes of selection, natural selection and somatic selection.

The result is a subtle and multilayered affair, full of loops and layers. From genes to proteins, from cells to orderly development, from electrical activity to neurotransmitter release, from sensory sheets to maps, from shape to function and behavior, from social communication back to any and all of these levels, we are confronted with a system of somatic selection that is continually subjected to natural selection (figure 14–1). Is it any wonder that philosophers, thinking about the problem of the mind without this knowledge, were tempted to postulate entities, that physicists have been tempted to postulate exotic new material fields, and that those in hope of immortality continue to postulate eternal spirits?

It may come as a disappointment to such thinkers that the answers to many of the fundamental problems of mind will come from analyzing the complexity of its organization, which is governed by novel ordering principles. But considered again, how rich, how full of surprise, how much

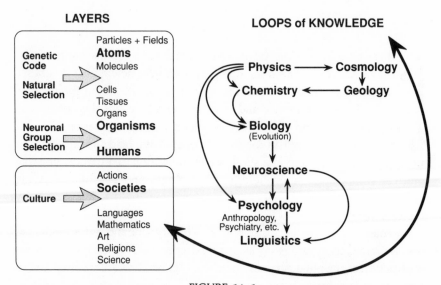

FIGURE 14–1

Layers of biological organization and loops of knowledge. See figure P–1 for the scales affecting the layers and loops.

more of a piece with the grand and compatible theories of evolution and physics this alternative is!

The untangling of the complexity has barely begun and the present effort at synthesis will undoubtedly seem paltry when it is over. But even at its early stages, the whole business of the matter of the mind requires a global view if we are to get anywhere. A simple analogical comparison will not do, nor will rationalism, nor will physics alone. A theory of the matter itself is needed. A major aim of the one described here is to provoke the construction by others of alternative theories within the same constraints—not philosophical hypotheses, not high-level formulations, but biological theories that challenge either the facts I have presented or the ways in which I have interpreted them.

In the meantime, I hope the reader will welcome a summary of the layers and the loops from my point of view. By now I assume that the strange vocabulary is familiar enough to be used here to sum up my position. I will proceed in reverse, from the theory adopted to my criticisms of the alternatives. Please keep in mind the generalization that as a selective system, the brain (especially the cerebral cortex) is a correlator. It correlates temporal inputs during its own development, and it correlates the properties of signals and scenes in its adult functioning to give rise to consciousness.

My first premise is that consciousness appeared as a result of natural selection. The mind depends on consciousness for its existence and functioning. A related notion is that consciousness is efficacious, enhancing fitness in certain environments. Consciousness arises from a special set of relationships between perception, concept formation, and memory. These psychological functions depend on categorization mechanisms in the brain. In addition, memory is influenced by evolutionarily established value systems and by homeostatic control systems characteristic of each species.

Primary consciousness is achieved by the reentry of a value-category memory to current ongoing perceptual categorizations that are carried out simultaneously in many modalities. It links parallel stimuli in time and space (including those not necessarily causally connected) into a correlated scene. In an individual animal, the features of that scene achieve salience from that animal's past values and learning history. Primary consciousness is limited to the remembered present. It is necessary for the emergence of higher-order consciousness, and it continues to operate in animals capable of higher-order consciousness.

Higher-order consciousness arises with the evolutionary onset of semantic capabilities, and it flowers with the accession of language and symbolic reference. Linguistic capabilities require a new kind of memory for the production and audition of the coarticulated sounds that were made possible by the evolution of a supralaryngeal space (see figure 12–1). The speech areas mediating categorization and memory for language interact with already evolved conceptual areas of the brain. Their proper function in a speech community connects phonology to semantics, using interactions with the conceptual areas of the brain to guide learning. This gives rise to a syntax when these same conceptual centers categorize the ordering events occurring during speech acts. As a syntax begins to be built and a sufficiently large lexicon is learned, the conceptual centers of the brain treat the symbols and their references and the imagery they evoke as an "independent" world to be further categorized. A conceptual explosion and ontological revolution—a world, not just an environment—are made possible by the interaction between conceptual and language centers.

By these means, concepts of self and of a past and a future emerge. Higher-order consciousness depends on building a self through affective intersubjective exchanges. These interactions—with parental figures, with grooming conspecifics, and with sexual partners—are of the same kind as those guiding semiotic exchange and language building. Affectively colored exchanges through symbols initiate semantic bootstrapping. The result is a model of a world rather than of an econiche, along with models of the past, present, and future. At the same time that higher-order consciousness frees us from the tyranny of the remembered present, however, primary consciousness coexists and interacts with the mechanisms of higher-order consciousness. Indeed, primary consciousness provides a strong driving force for higher-order processes. We live on several levels at once.

The Jamesian properties of these conscious processes (see chapter 11) depend on the function of the cerebral cortex and its appendages. The latter constitute the organs of succession—the cerebellum for smooth movement, the hippocampus for laying down long-term memory, and the basal ganglia for choosing motor patterns and attentional plans. Their functioning depends on the motion and action of the organism exploring its environment.

The resulting properties of subjectivity, intentionality, continuity, and change occur together in an apparent unity. These properties can be explained by the extended TNGS, requiring no assumptions beyond those of developmental selection, experimental selection, and reentry. New functions, including consciousness, are made possible by new evolutionary morphology connected in new ways to existing brain structures.

"Objective" science and language both depend on the metastability or constancy of objects in the physical world. The consciousness theory assumes that physics and evolution, supplemented by the assumptions of the TNGS, are sufficient to construct a science of mind. No scientific theory of a single actual mind is possible, however, any more than a scientific account of all historical events in the world is possible.

To remain scientific, the extended TNGS must assume that both the human subject and the human scientific observer who studies that subject experience qualia. This assumption is necessary to assure that meaningful intersubjective scientific exchange occurs. According to the theory, qualia are categorizations by higher-order consciousness of the "scenes" and "memories" provided by primary consciousness. They involve recategorical relationships that are ultimately governed by how evolutionarily selected values interact with memory.

Creatures that have primary consciousness alone can neither report qualia nor reflect on them. If they experience them (and we can only infer that they do), they experience them solely in the remembered present. On the basis of morphological comparisons, we may conceive of three levels of sensory properties in the evolution of animals with neurons:

1. Responses to stimuli with aversive and consummatory responses directly governed by selection for evolutionary values. An example is the lobster, capable of learning and long-term memory but not of primary consciousness.

2. Stimuli eliciting responses in animals having primary consciousness. Mental life consisting of Jamesian scenes correlating value and perceptual categorization, but no socially constructed self. Anatomical bases for qualia and their discrimination according to different modalities. No categorization of qualia over time by a *subject*, but long-term memory (nonconscious as such) based on qualia in the remembered present. Example: dogs.

3. Stimuli with aversive and appetitive significance transformed by animals having higher-order consciousness into a world, not just an econiche. Full-blown qualia capable of being refined, remembered, altered, and reported, as in wine tasting. Example: humans. Extreme example: sainthood, with denial of all biological imperatives, including unusual responses to painful qualia, on the basis of deeply held belief.

For now, we can only speculate on such matters. But we do know that higher-order consciousness leads to the construction of an imaginative domain, one of feeling, emotion, thought, fantasy, self, and will. It constructs artificial objects that are mental. In culture, these acts lead to studies of stable relations among things (science), of stable relations among stable mental objects (mathematics), and of stable relations between sentences that are applicable to things and to mental objects (logic). One possible reason for the incompleteness of such domains, as shown for mathematics by Kurt Gödel, is that pattern formation in the mind always requires the higher-order bootstraps that are necessary for consciousness. Thinking occurs in terms of *synthesized* patterns, not logic, and for this reason, it may always exceed in its reach syntactical, or mechanical, relationships.

The analogy between the mind and a computer fails for many reasons. The brain is constructed by principles that assure diversity and degeneracy. Unlike a computer, it has no replicative memory. It is historical and value driven. It forms categories by internal criteria and by constraints acting at many scales, not by means of a syntactically constructed program. The world with which the brain interacts is not unequivocally made up of classical categories. (It is true, however, that some "natural objects" appear to follow these categories because of the interactive features of our phenotype and the physical properties of these objects.)

The world, therefore, is not like a piece of computer tape. Physics, which studies such a world, describes its formal correlative properties but does not contain a theory of unique categories for the partitioning of macroscopic objects. As I point out in the Postscript, objectivism fails.

Categorization mechanisms work through global mappings that necessarily involve our bodies and our personal history. Perception is therefore not *necessarily* veridical (see, for example, the Kanizsa triangle in figure 4–2). In our behavior we are driven by a recategorical memory under the influence of dynamic changes of value. Beliefs and concepts are individuated only by reference to an open-ended environment, the description of which cannot be specified in advance. Our modes of categorization and the use of metaphor in our thinking (mapping one thing to another in a different domain) reflect these observations.

I argue at length in the Postscript that the cognitive science view of the mind based on computational or algorithmic representations is ill-founded. Mental representations that are supposedly syntactically organized (in a "language of thought") and then mapped onto a vaguely specified semantic model or onto an overly constrained objectivist one are incompatible with the facts of evolution. The properties proposed by these cognitive models are incompatible with the properties of brains, bodies, and the world. The extended TNGS purports to explain how embodiment of mind takes place and thus connects cognition to biology. It provides a consistent basis for explaining how meaning arises from embodiment as a result of referential interactions. A rich field of study concerned with exactly how our concepts map onto our bodies is presently in its earliest stages.

Why have I rejected as a basis for mind the apparent elegance of axiomatic and syntactic systems? Axiomatic systems often *seem* to provide the right clue as to how the mind works, especially when taken together with physics. But they are social constructions that are the *results* of thought, not the basis of thought. Their roots lie in the mathematical logic of the nineteenth century. They flowered with David Hilbert, were modulated and circumscribed by Gödel, and are often conceived of in a typological or essentialist fashion. They are *not* a good model for the mind, for the mind must *preexist* to create and drive them. Consciousness is essential for their formulation and also for the Platonism that they sometimes inspire, but the facts show that consciousness arose by evolutionary, not typological, means. Darwin was right: Morphology led to mind, and on this issue Wallace, who felt that natural selection could not explain the human mind, was wrong. Plato is not *even* wrong; he is simply out of the question.

It may be useful here to mention the obvious—that the evolution of consciousness depended on certain temperatures. The stability of any physical object that science and the conscious scientist describe is the same stability that "glues together" a naming event. There could be no consciousness at 10^6°C. It emerged at a certain time, place, and at a much lower temperature, one that allowed chemistry to occur. To say so is to reject panpsychism as a theory of mind. (I discuss this matter further in chapter 20, which considers the ultimate origins of mind.)

Consciousness is central to human behavior, society, language, and science. Imagine the opposite and you have to postulate a prescribed world tape, a "brain-computer," and a very boring "world programmer." The TNGS, with its complexity of layers and loops, appears to be more in line with the facts of biology and seems preferable because it fits much more of our own experience.

With this statement of an obvious personal preference, I turn to the final part of this book. I have entitled it *Harmonies* to underscore the fruitful interactions that a science of mind must have with philosophy, medicine, and physics. These fields are all different, but truly interesting harmonies are based on the consonance of different entities, not on identity or unison. We all hope for a resolution of conflicting visions, for clarification of thought, and for harmonies between ideas. I am no exception and I have no intention of missing the chance to philosophize, first, about philosophy itself; second, about the idea of the self, its thoughts and its disorders; third, about the possibility of making conscious artifacts; and last, about the grand themes of a future science that will reveal more clearly the connection between physics and psychology.

In *Modes of Thought*, Whitehead pointed out that philosophy is the attempt to make manifest the fundamental evidence as to the nature of things. In the same work, he remarked that scientific reasoning is completely dominated by the presupposition that mental functionings are not properly part of nature. He deplored this and hoped that a proper connection between the mental and physical could be forged within science itself. That was in 1933. Now, in the 1990s, a glimpse into how that could come about may be possible without closing the door on philosophy, which above all is an attitude of mind.

PART IV

HARMONIES

This final set of chapters asks about the implications of our new brain theory for human (and some inhuman) concerns. It pleads for an open-mindedness about the mind. It suggests that our knowledge is not incorrigible, that we are deeply embedded in the matter of the world as well as in the matter of the mind, that we are each of us unique as individuals (and importantly so), that our thinking in a culture is a critical matter for our being human and for our grasping of meaning, and that, even in disease, our minds are marvelously adaptive. It also suggests that the time is not hopelessly remote when we may be able to build artifacts that share some of our own psychological properties.

Above all, it suggests that constructing an adequate theory of the brain promises to offer bases for new harmonies, including those according to which we may place ourselves in the universe. In the final chapter, I attempt to answer the question: If one were to name two grand scientific ideas or concepts that together capture how we may ground ourselves and help determine where we are in the order of things, what would those ideas be?

CHAPTER 15

A Graveyard of Isms: Philosophy and Its Claims

Any two philosophers can tell each other all they know in two hours.

I don't see why a man should despair because he doesn't see a beard on his Cosmos. If he believes that he is inside of it, not it inside him, he knows that consciousness, purpose, significance, and ideals are among its possibilities . . . and the business of philosophy is to show that we are not fools for doing what we want to do.
—Oliver Wendell Holmes, Jr.

A concern with the mind and its workings has permeated philosophy from its beginnings. In his review of the concepts of philosophy, Arthur Danto defines almost all philosophical positions in terms of what he calls a basic cognitive episode. This notion harks back to Descartes and is expressed as a relationship between three components: a subject, a representation, and the world. The relationship between the world and the subject is that of causality. The relationship between the world and representation is that of truth, and the relationship between the subject and representation is that which the subject has with him- or herself. Danto calls humans representing beings or representational beings, and I believe he falls into the trap we have already warned against when he advances the view that the body is "sententially structured," meaning that the task of the neurosciences is "to show how nervous tissue represents." If one does not hold him too closely to this view, which skirts dangerously close to ideas of coding and instruction, his triad is a useful one, for it can help us see how, as individuals, *we* (not our brains) "represent" the world.

To a scientist, philosophy can be a disconcerting business. Science is supposed to provide a description of the laws of that world and of how they may be applied. Philosophy by contrast, has no proper subject matter of its own. Instead, it scrutinizes other areas of knowledge for clarity and consistency. Furthermore, unlike science, it may be called immodest. There is no partial philosophy; it is complete with each philosopher. Like a child exploding into a grasp of language, the philosopher must not simply describe an environment but construct a whole world. Each time a philosophical construction is attempted, there is a world view behind it, and a personal one at that. The entitling of such world views as various "isms" (table 15–1) leads to an interesting collection of ways in which Danto's triad has been dissected with respect to the importance accorded its internal relations. I will not discuss them all here; the shallow account given in chapter 4 should give you a feeling for a few of them. Readers might want to amuse themselves with a dictionary or an encyclopedia of philosophy to check their ramifications. The trouble is that each "ism" is likely to spell the rejection of the last, as each philosopher constructs a unique point of view. Philosophy is a graveyard of "isms."

Why bother with it then? Because philosophy attempts to apply thought to *all* aspects of our individual and collective existence; because its history

TABLE 15–1
Some Philosophical "Isms"*

Empiricism	Monism	Realism
Rationalism	Dualism	Idealism
Phenomenalism	Pluralism	Foundationalism
Reductionism	Epiphenomenalism	Essentialism
Objectivism	Materialism	Behaviorism
Operationalism	Panpsychism	(Philosophical behaviorism)
Instrumentalism	Determinism	Representationalism
Logical positivism	Compatibilism	Functionalism
Foundationalism	Incompatibilism	Interactionism
Pragmatism	Occasionalism	Internalism
Evolutionism		Externalism
Selectionism		Existentialism

This list might go on beyond all decent bounds were one to include moral, aesthetic, clinical, religious, and political ideologies. Isms most definitely ruled out by scientific study include geocentrism, vitalism, and mechanism. Of course, not all doctrines are isms (and possibly vice versa) but they could be made to be, and that is the danger.

is closely intertwined with that of psychology; and because a new scientific view of the mind based on biology may help give philosophy a new lease on life.

An anonymous university president is quoted by John Barrow and Frank Tipler in their book *The Anthropic Cosmological Principle:*

> Why is it that you physicists always require so much expensive equipment? Now, the Department of Mathematics requires nothing but money for paper, pencils, and waste paper baskets, and the Department of Philosophy is better still. It doesn't even ask for waste paper baskets.

Contrast this with Einstein, who remarked that the theoretical physicist's most important tool is the wastebasket!

Philosophy in this century has been characterized by a retreat from the grand synthetic goals of its past, goals touched on in chapter 4. Since Wittgenstein, a good part of philosophy has been concerned with tidying up logic and language. Since Edmund Husserl, another part has been concerned with a deliberately nonscientific set of reflections on consciousness and existence, or phenomenology, as it is called. It is certainly worth asking whether a biologically based theory of mind would invigorate these areas of thought and perhaps even give philosophy a new turn.

Let us join the game and see how many isms must fall if we take a scientific position on the mind. Of course, this means saying something about the limits of science and of knowledge itself. We should first state the assumptions of a scientific view:

1. There is a real world—one described by the laws of physics, which apply everywhere. (This is the physics assumption.)
2. We are embedded in that world, follow its laws, and have evolved from an ancient origin. The mind arose on the basis of new evolutionary morphology. (This is the evolutionary assumption.)
3. It is possible to put the mind back into nature. A science of mind based on biology is feasible. The way to avoid vicious circles and dead ends is to construct a brain theory based on selectionist principles. (This is the central argument of this book.)

If we accept these assumptions and the previous arguments of this book, we can immediately add to the graveyard. Dualism, panpsychism, epiphenomenalism, idealism, representationalism, empiricism, and essentialism are all incompatible with both the assumptions listed above and the evidence from psychology and neuroscience, as well as biology itself. I shall not belabor the arguments but simply state, in the same sequence as shown in this list, the following presumably lethal sentences: There is no *res cogitans*; particles are not conscious; consciousness is evolutionarily efficacious; the world exists and persists independent of mind and preexisted before its appearance; the brain is a selective system and not a Turing machine; sense data are not the basis of the mind; the "world" does not consist of classical categories; typology is destroyed by biology. And over the last 300 years, science has already destroyed the more parochial ideas of geocentrism, vitalism, and simple mechanism.

So much for destruction, at least for the moment. What can we say constructively about science and about the possibility of a theory of knowledge based on biology—a biologically based epistemology?

To construct a reasonable account, we must recognize that modern particle physics and field theory have eliminated the notion of the world as a deterministic or clockwork mechanism. This does not mean that mechanisms cannot be described or be useful (as they are to both macroscopic physics and biology). It means simply that the universe cannot be sensibly considered *at all scales* in such terms (see figure P–1). We must also recognize that a death blow was dealt to essentialism and Platonism by the Darwinian theory of evolution. Finally, we must take into account the fact that systems of natural selection in evolution (which are historical systems) gave rise to somatic selective systems capable of dealing with novelty within an individual's lifetime.

This last point is not so securely based—the idea that there are sciences of recognition, of which neuroscience is a central one has not yet been generally accepted for studies of the brain. But if we assume that the main premises of neural Darwinism are correct (and evidence in support of them is mounting), several interesting conclusions may be drawn. First of all, we need not reach beyond biology itself to mount any exotic explanations of the mind. Remember that the assumptions of the TNGS—developmental selection and variance, synaptic selection, and differential amplification within reentrant systems—are *all* of the principles proposed by that theory. No new principles need be adduced to account for consciousness—only new evolutionary morphologies. Second, these notions, if correct, rule out a *general* description of the workings of the brain as a Turing machine or computer. And third, while substance dualism (the Cartesian variety) and property dualism (the notion that psychology can be satisfactorily described only in its own terms) are ruled out, we must admit to a distinction between selective and nonselective material systems. This distinction identifies living and mental systems as selective. In other words, there is a real distinction between biology (or psychology) and physics. While admitting that the laws of physics apply to both intentional and nonintentional systems, this position at the same time denies that fancy physics—such as quantum gravity or other specialized concepts of fundamental physics—are required to explain mind.

Where, after this, do we stand with the isms?

By taking the position of biologically based epistemology, we are in some sense realists and also sophisticated materialists. Given the facts of development and evolution, we deny teleology (the doctrine of final causes or ultimate goals). But at the same time, we admit that evolution *can* select animals in such a way that they have general goals, purposes, and values, so that they embody what have been called teleonomic systems. As Mayr put it, teleonomy is a prediction of the past. The past experience of natural selection adjusts the set points of value systems (for example, those related to hunger, thirst, sexual responses) that are adaptive for survival. In our case, the brain of a conscious human being, serving as a somatic selective system, uses value constraints to project the future in terms of categories and goals. And by assuming that the brain is a somatic selective system, we rule out the idea of the little man or homunculus in the head. He is no more necessary to the sciences of somatic recognition than special creation or the argument from design is to evolution. If he is *res cogitans*, he is exorcised.

Mind, which arose from material systems and yet can serve goals and purposes, is nevertheless a product of historical processes and of value-based constraints related to evolution. What bounds does this place on our knowledge and our freedom?

That consciousness arose in the material order does not restrain intellectual trade; philosophy itself is witness to this conclusion. But it does limit us, despite our capacity to extend our senses and our powers of calculation through physical devices. Given how meaning is defined in this book, we must accept a position of *qualified* realism. Our description of the world is qualified by the way in which our concepts arise. And although there may be infinite freedom *within* a grammar, our language and our ideas of meaning go far beyond the rules of grammar. I have already described how I think meaning arises from embodiment through neuronal group selection and reentry. Despite the remarkable extensions of meaning by our calculations and our experiments, we must admit that we may well be limited in our thought by the way in which we are constituted as products of evolutionary morphology.

We can add three more important elements to this picture of qualified realism and biologically based epistemology. These are: (1) an extraordinary density of real-world events exists; even given our ability to categorize a large number of them, we can hardly exhaust their description; (2) many events are irreversible; and (3) in each individual, sensation and perception follow unique, irreversible, and idiosyncratic courses.

According to biologically based epistemology and qualified realism, knowledge *must* remain fragmentary and corrigible. There is no Cartesian certainty. But, you may object, what of mathematical certainty, of analytic relations and tautologies? This is no place for an extended discussion of these matters. It is useful, however, to make it clear that such systems are artificial ones, created by the mind through social interactions and individual manipulations of symbols. The most basic of these systems, arithmetic, has been shown by Gödel to be incomplete. I would characterize the study of mathematics, as Philip Davis and Reuben Hersh aptly put it in *The Mathematical Experience*, as the study of stable or invariant mental objects. Although our subtle materialism hardly asserts that these relations are not meaningful, it denies them a separate Platonic existence.

We have already described some other limits. For example, given the limitations on our knowledge and our "locked-in" state, the exact idiosyncratic and irreversible path of an individual's qualia is not accessible to another individual. (Indeed, it can become inaccessible even to its owner). Moreover, we must accept that death means the irrevocable loss of an individual and of that individual's being. Death is not an experiment: There is nothing to report. Minds do not exist disembodied. These are obvious limits to the knowledge of persons and to science. What we are trying to do, nonetheless, is place ourselves clearly within the world view given by science.

In addition to qualifying our realism, we must consider questions of history and culture and ones related to value and purpose. This may seem strange in a discussion of science, which is supposed to be value-free. But the science touted as value-free is that based on the Galilean position, a physical science that quite deliberately and justifiably removed the mind from nature. A biologically based epistemology has no such luxury.

It is worth dwelling on this matter a bit, for it reveals much about the place of science in our lives. Conscious human experience has given rise to culture, and culture to history. History is not simply a chronicle but an interpretation, encompassing suspected causes and values. Science has emerged within history, and it attempts to describe, with considerably more certainty, the boundaries of the world—its constraints and its physical laws. But these laws cannot replace history or the actual courses of individual lives. A set of laws is not a substitute for experience and it is certainly not equivalent to a set of events. Laws do not and cannot exhaust experience or replace history or the events that occur in the actual courses of individual lives. Events are denser than any possible scientific description. They are also microscopically indeterminate, and, given our theory, they are even to some extent macroscopically so.

It may seem that I am attempting to limit *a priori* the capabilities of scientific description. Nothing of the sort. I am simply pointing out that, even if science succeeds in putting the mind back into nature, it will not, according to the description we have given, be able to describe individual or historical experience adequately. But it does provide a satisfactory (indeed, the best) description of the *constraints* on experience.

What are some of those constraints when viewed through a selectionist theory of the mind? One comes out of the extended TNGS: No selectionally based system works value-free. Values are necessary constraints on the adaptive workings of a species. In our species, the commonalities of physiological function, hunger, and sex imply a set of mutually shared properties. The brain is structured so as to play a key role in regulating the evolutionarily derived value systems that underlie these properties. Undoubtedly, these value systems also underlie the higher-order constructions that make up individual aims and purposes. We categorize on value.

Once higher-order consciousness arose, values at the biological level could be modified, though only to a degree. Of course, as mentioned earlier, we must admit the possibility of an almost total denial of biological values on the part of those organisms we call martyrs and saints. Only creatures endowed with higher-order consciousness can so transcend the dictates of biology. If one agrees to omit saints from consideration, the insertion of aims and purposes and ethical values into social systems, however far they are from basic biological value systems, almost certainly results from the original need for value in guiding the selectional systems of the brain. In any culture, decisions involving social value must come before those that elevate the interests of science, however important scientific knowledge may be.

Science has turned out to be eminently practical, as it must be, given its service to the verifiable truth. Modern society and its economics depend increasingly on scientific technology, and scientific beliefs are being assimilated by increasing numbers of people. In addition to curiosity, however, greed now often drives the search for knowledge. The good to be derived from that search, however motivated, is that science may usefully transform our material conditions of being (provided we remain clear about values). But a constant tension remains in balancing private and public good, as the history of this industrial and atomic century has dramatically taught us. Power is not insight, and the shift from a science based solely on physics to one based on physics and biology is likely to lend us deep insight and to help change how important social decisions are made. A biologically based epistemology will have valuable things to say about such decisions as we discover more about our brains.

Knowing our place in the world on the basis of a biologically based theory of the mind will also reveal our limits and restrain our philosophical ambition. But in certain directions the limits are hardly very constraining. The imagination of conscious humans in culture is potentially limitless. Constrained as we are (each of us locked into our own conscious experience), mortal as we may be, and qualified as our realism is, the future remains open; it is not predetermined. We do not have the security of foundationalism or a first philosophy, nor the ability to know with certainty all that we can appreciate or place into a pattern. Most of us cannot deny our evolutionarily selected biological values, nor should we, given that they provide a common ground for our moral decisions. But the history of scientific discovery and the achievements of human imagination promise constant surprise and, with the rise of brain science, provide an increasingly solid basis for attempts to place ourselves within our own world description.

In the end, therefore, we must conclude that we have not been able to kill all the isms. We have suggested a favored set: qualified realism, sophisticated materialism, selectionism, and Darwinism. Indeed, considering their significance and relating them to what physics and biology together have to offer should enrich philosophy and ensure its harmony with science. After all, thinking is not the same as a theory of mind, and there is much thinking to be done about selective systems. By its very nature, the position on biologically based epistemology that I have taken here implies that science-free phenomenology and grammatical exercises, whatever their value, place too narrow a set of limits on the philosophical enterprise. Philosophy needs a new turn.

I believe that neuroscience will play a central role in such a development. But it will not be a development in which a simple reductionism to quantum fields, to strange particles or the like has any sway. It will instead be one in which the task is to see how selectional systems of the brain grounded on value give rise to meaning and selfhood and how the self construes the boundaries of the world. This task feeds back onto physics and forward onto social views of the worth of the individual. Let us explore a few of its implications.

CHAPTER 16

Memory and the Individual Soul: Against Silly Reductionism

Science cannot solve the ultimate mystery of Nature. And it is because in the last analysis we ourselves are part of the mystery we are trying to solve.

—Max Planck

If I had to live over again, I'd live over a delicatessen.

—Woody Allen

From the last quarter of the seventeenth century to the last decade of the eighteenth, an explosion of creativity called the Enlightenment changed the history of ideas. Its reigning views were many, but above all it was dedicated to reason, to science, and to human freedom and individuality. Its underlying science was physics, the system of Newton, and its philosophy of society was, in large measure, that of Locke. Yet the Enlightenment ideas of causality and determinism, along with its mechanistic view of science, undermined hopes for a theory of human action based on freedom. If we are determined by natural forces—by mechanism—we cannot easily put together a consistent picture in which a free individual makes moral choices. Moreover, while the ideas of the Enlightenment paid much attention to the role of reason and culture in such choices, there was no general notion of how deeply the minds of all humans (including those of "reasonable" human beings—that is, the "cultured") were influenced by unconscious forces and by emotion.

Whatever forms it took at various times and places, the overriding Enlightenment view was a secular one that forged many of the ideas underlying modern democracy. But despite its valuable heritage, the Enlightenment is over. The first great blow to its ideas came with Hume's damaging attacks on both rationalism and the notion of human progress as linked to natural science. Its major fault was its inability to create an adequate scientific description of a human individual to accompany its description of a machinelike universe. Its social failure was its inability to go beyond the concept of a society composed of self-seeking, commercially successful individuals with a shallow view of "humanism." Certainly, Enlightenment thinkers attempted to provide us with a larger, more inspiring view of ourselves. But its science was a mechanistic physics and it had no body of data or ideas with which to link the world, the mind, and society in the style of scientific reason to which it aspired. Whatever the Enlightenment's failures and inconsistencies, however, it left us with high hopes for the place of the individual in society.

Can we expect to do better with a sound scientific view of mind? In this chapter I hope to show that the kind of reductionism that doomed the thinkers of the Enlightenment is confuted by evidence that has emerged both from modern neuroscience and from modern physics. I have argued that a person is not explainable in molecular, field theoretical, or physiological terms alone. To reduce a theory of an individual's behavior to a theory of molecular interactions is simply silly, a point made clear when one considers how many different levels of physical, biological, and social interactions must be put into place before higher-order consciousness emerges. The brain is made up of 10^{11} cells with at least 10^{15} connections. Each cell has a fantastically intricate regulatory biochemistry constrained by particular sets of genes. These cells come together during morphogenesis and exchange signals in a place-dependent fashion to make a body and a brain with enormous numbers of control loops, all obeying the homeostatic mechanisms that govern survival. Selection on neuronal repertoires leads to changes in myriad synapses as cells die or differentiate. An animal's survival and motion in the world allow perceptual and conceptual categorization to occur continually in global mappings. Memory dynamically interacts with perceptual categorization by reentry. Learning involving the connection of categorization to value (in its most subtle form within a speech community) links symbolic and semantic abilities to conceptual centers that already provide embodied structures for the building of meaning.

A calculation of the significant molecular combinations of such a sequence of events, even in identical twins, is almost impossible, and in any case, useless. The mappings are many—many, and the processes are individual and irreversible. I wonder what Enlightenment humanists would have made of all this. Diderot, who as we saw in chapter 3 speculated about the nervous system of his friend in *Le Rêve de d'Alembert*, might have been pleased. Diderot's view of human consciousness opened up the possibility that to be human was to go beyond mere physics.

I have taken the position that there can be no complete science, and certainly no science of human beings, until consciousness is explained in biological terms. Given our view of higher-order consciousness, this also means an account that explains the bases of how we attain personhood or selfhood. By selfhood I mean not just the individuality that emerges from genetics or immunology, but the personal individuality that emerges from developmental and social interactions.

Selfhood is of critical philosophical importance. Some of the problems related to it may be sharpened by the selectionist view I have taken on the matter of mind. Please remember, however, that no scientific theory of an individual self can be given (our qualia assumption). Nonetheless, I believe that we can progress toward a more complete notion of the free individual, a notion that is essential to any philosophical theory concerned with human values.

The issues I want to deal with are concerned with the relationship between consciousness and time, with the individual and the historical aspects of memory, and with whether our view of the thinking conscious subject alters our notion of causality. I also want to discuss briefly the connection between emotions and our ideas of embodied meaning. All of these issues ultimately bear upon the matter of free will and therefore upon morality under mortal conditions.

According to the extended TNGS, memory is the key element in consciousness, which is bound up with continuity and different time scales. There is a definite temporal element in perceptual categorization, and a more extended one in setting up a conceptually based memory. The physical movements of an animal drive its perceptual categorization, and the creation of its long-term memory depends on temporal transactions in its hippocampus. As we have seen, the Jamesian properties of consciousness may be derived from the workings of such elements. But in human beings, primary consciousness and higher-order consciousness coexist, and they each have different relations to time. The sense of time past in higher-order consciousness is a *conceptual* matter, having to do with previous orderings of categories in relation to an immediate present driven by primary consciousness. Higher-order consciousness is based not on ongoing experience, as is primary consciousness, but on the ability to model the past and the future. At whatever scale, the sense of time is first and foremost a conscious event.

The ideas of consciousness and "experienced" time are therefore closely intertwined. It is revealing to compare the definition of William James, who stated that consciousness is something the meaning of which "we know as long as no one asks us to define it," with the reflections of St. Augustine, who wrote in his *Confessions*, "What then is time? If no one asks me, I know what it is. If I wish to explain to him who asks me, I do not know." The notion of continuity in personal, historical, and institutional time was a central one in Augustine's thought.

Time involves succession. An intriguing suggestion about the connection between time and the idea of numbers has come from L. E. J. Brouwer, a proponent of intuitionism in mathematics. He suggests that all mathematical elements (and particularly the sequence of natural numbers) come from what he calls "two-icity." Two-icity is the contrast between ongoing conscious experience (with primary consciousness as a large element) and the direct awareness of past experience (requiring higher-order consciousness). What is intriguing about this is that it suggests that one's concept of a number may arise not simply from perceiving sets of things in the outside world. It may also come from inside—from the intuition of twoness or two-icity plus continuity. By recursion, one may come to the notion of natural numbers.

Whatever the origins of such abstractions, the personal sense of the sacred, the sense of mystery, and the sense of ordering and continuity all have connections to temporal continuity as we experience it. We experience it as individuals, each in a somewhat different way.

Indeed, the flux of categorization, whether in primary or higher-order consciousness, is an individual and irreversible one. It is a history. Memory grows in one direction; with verbal means, the sense of duration is yet another form of categorization. This view of time is distinguishable from the relativistic notion of clock time used by physicists, which is, in the microscopic sense, reversible. Aside from the variation and irreversibility of *macroscopic* physical events recognized by physicists, a deep reason for the irreversibility of individually experienced time lies in the nature of selective systems. In such systems, the emergence of pattern is *ex post facto*. Given the diversity of the repertoires of the brain, it is extremely unlikely that any two selective events, even apparently identical ones, would have identical consequences. Each individual is not only subject, like all material systems, to the second law of thermodynamics, but also to a multilayered set of irreversible selectional events in his or her perception and memory. Indeed, selective systems are by their nature irreversible.

This "double exposure" of a person—to real-world alterations affecting nonintentional objects as well as to individual historical alterations in his or her memory as an intentional subject—has important consequences. The flux of categorizations in a selective system leading to memory and consciousness alters the ordinary relations of causation as described by physicists. A person, like a thing, exists on a world line in four-dimensional spacetime. But because individual human beings have intentionality, memory, and consciousness, they can sample patterns at one point on that line and on the basis of their personal histories subject them to plans at other points on that world line. They can then enact these plans, altering the causal relations of objects in a definite way according to the structures of their memories. It is as if one piece of spacetime could slip and map onto another piece. The difference, of course, is that the entire transaction does not involve any unusual piece of physics, but simply the ability to categorize, memorize, and form plans according to a conceptual model. Such an historical alteration of causal chains could not occur in so rich a way in any combination of inanimate nonintentional objects, for they lack the appropriate kind of memory. This is an important point in discriminating biology from physics, an issue I discuss further in chapter 20.

In certain memorial systems, unique historical events at one scale have causal significance at a very different scale. If the sequence of an ancient ancestor's genetic code was altered as a result of that ancestor's travels through a swamp (driven, say, by climatic fluctuations), the altered order of nucleotides, if it contributed to fitness, could influence present-day selectional events and animal function. Yet the physical laws governing the actual *chemical* interaction of the genetic elements making up the code (the nucleotides) are deterministic. No deterministic laws at the chemical level could alone, however, explain the *sustained* code change that was initiated and then stabilized over long periods as a result of complex selectional events on whole animals in unique environments.

Memorial events in brains undergoing selectional events are of the same ilk. Because the environment being categorized is full of novelty, because selection is *ex post facto*, and because selection occurs on richly varied historical repertoires in which different structures can produce the same result, many degrees of freedom exist. We may safely conclude that, in a multilevel conscious system, there are even greater degrees of freedom. These observations argue that, for systems that categorize in the manner that brains do, there is macroscopic indeterminacy. Moreover, given our previous arguments about the effects of memory on causality, consciousness permits "time slippage" with planning, and this changes how events come into being.

Even given the success of reductionism in physics, chemistry, and molecular biology, it nonetheless becomes silly reductionism when it is applied exclusively to the matter of the mind. The workings of the mind go beyond Newtonian causation. The workings of higher-order memories go beyond the description of temporal succession in physics. Finally, individual selfhood in society is to some extent an historical accident.

These conclusions bear on the classical riddle of free will and the notion of "soft determinism," or compatibilism, as it was called by James Mill. If what I have said is correct, a human being has a degree of free will. That freedom is not radical, however, and it is curtailed by a number of internal and external events and constraints. This view does not deny the influence of the unconscious on behavior, nor does it underestimate how small biochemical changes or early events can critically shape an individual's development. But it does claim that the strong psychological determinism proposed by Freud does not hold. At the very least, our freedom is in our grammar.

These reflections, and the relationship of our model of consciousness to evolved values bear also on our notion of meaning. Meaning takes shape in terms of concepts that depend on categorizations based on value. It grows with the history of remembered body sensations and mental images. The mixture of events is individual and, in large measure, unpredictable. When, in society, linguistic and semantic capabilities arise and sentences involving metaphor are linked to thought, the capability to create new models of the world grows at an explosive rate. But one must remember that, because of its linkage to value and to the concept of self, this system of meaning is almost never free of affect; it is charged with emotions. This is not the place to discuss emotions, the most complex of mental objects, nor can I dedicate much space to thinking itself. I consider them in the next chapter. But it is useful to mention them here in connection with our discussion of free will and meaning. As philosophers and psychologists have often remarked, the range of human freedom is restricted by the inability of an individual to separate the consequences of thought and emotion.

Human individuals, created through a most improbable sequence of events and severely constrained by their history and morphology, can still indulge in extraordinary imaginative freedom. They are obviously of a different order than nonintentional objects. They are able to refer to the world in a variety of ways. They may imagine plans, propose hopes for the future, and causally affect world events by choice. They are linked in many ways, accidental and otherwise, to their parents, their society, and the past. They possess "selfhood," shored up by emotions and higher-order consciousness. And they are tragic, insofar as they can imagine their own extinction.

Often it is said that modern humans have suffered irreversible losses from several episodes of decentration, beginning with the destruction of earlier cosmologies placing human beings at the center of the universe. The first episode, according to Freud, however, took place when geocentrism was displaced by heliocentrism. The second was when Darwin pointed out the descent of human beings. And the third occurred when the unconscious was shown to have powerful effects on behavior. Well before Darwin and Freud, however, the vision of a Newtonian universe led to a severe fatalism, a view crippling to the societal hopes of Enlightenment thought. Yet we can now see that if new ideas of brain function and consciousness are correct, this fatalistic view is not necessarily justified. The present is not pregnant with a fixed programmed future, and the program is not in our heads. The theories of modern physics and the findings of neuroscience rule out not only a machine model of the world but also such a model of the brain.

We may well hope that if sufficiently general ideas synthesizing the discoveries that emerge from neuroscience are put forth, they may contribute to a second Enlightenment. If such a second coming occurs, its major scientific underpinning will be neuroscience, not physics.

The problem then will be not the existence of souls, for it is clear that each individual person is like no other and is not a machine. The problem will be to accept that individual minds are mortal. Given the secular views of our time, inherited from the first Enlightenment, how can we maintain morality under mortal conditions? Under present machine models of the mind this is a problem of major proportions, for under such models it is easy to reject a human being or to exploit a person as simply another machine. Mechanism now lives next to fanaticism: Societies are in the hands either of the commercially powerful but spiritually empty or, to a lesser extent, in the hands of fanatical zealots under the sway of unscientific myths and emotion. Perhaps when we understand and accept a scientific view of how our mind emerges in the world, a richer view of our nature and more lenient myths will serve us.

How would humankind be affected by beliefs in a brain-based view of how we perceive and are made aware? What would be the result of accepting the ideas that each individual's "spirit" is truly embodied; that it is precious *because* it is mortal and unpredictable in its creativity; that we must take a skeptical view of how much we can know; that understanding the psychic development of the young is crucial; that imagination and tolerance are linked; that we are at least all brothers and sisters at the level of evolutionary values; that while moral problems are universal, individual

171

instances are necessarily solved, if at all, only by taking local history into account? Can a persuasive morality be established under mortal conditions? This is one of the largest challenges of our time.

What will remain unclear until neuroscience grows more mature is how any of these issues can be linked to our history as individuals in a still-evolving species. In any case, silly reductionism and simple mechanism are out. A theory of action based on the notion of human freedom—just what was missing in the days of the Enlightenment—appears to be receiving more and more support from the scientific facts. We may now examine the connection of these facts to thought itself.

CHAPTER 17

Higher Products: Thoughts, Judgments, Emotions

There is in us something wiser than our head.
—Arthur Schopenhauer

How can a book on the matter of the mind pay so little attention to thinking, willing, and judging, or to feeling, emotion, and dreaming? Partly, this has to do with my original intentions, which were to describe the necessary bases for consciousness and meaning in a scientific fashion. I have attempted this in the faith that further and more sufficient psychological explorations can be launched once this description is substantiated. To pursue any one of these higher products of the mind's working would require a separate book. Nevertheless, I want to comment here on how our theses about the mind may be connected to psychological activities.

Consciousness is considered by some to be the same as thinking. I think this is too crude an identification, for thought has additional acquired components: a complex of images, intentions, guesses, and logical reasoning. It is a mixture of several levels of mental activity. At its highest and most abstract reaches, it is a skill, one that depends on symbolic abilities. With the exception of the spatial abilities exhibited in artistic thinking and the tonal and rhythmic activities of musical thinking, higher thought depends strongly on both language and logic, on an inner dialogue between the thinker and another "interlocutor" of whose existence the thinker may not be aware. This is the "two in one" to which Hannah Arendt refers in her book *The Life of the Mind.* She points out the distinction in German between *Vernunft*, pure thought or reason, and *Verstand*, understanding with a direct connection to the cognitive processes of perception, feeling, and the like.

I am not sure this distinction is useful in scientific terms but it does serve to emphasize how far thought can go. The thinker in the mode of pure thought is so immersed in a specific attentive state related to the project of thought that he or she is truly "abstracted"—unaware of time, space, self, and perceptual experience. One may say that in the pursuit of these levels of meaning and abstraction, "thought is nowhere." But this is simply a metaphor to express the individual's degree of removal from awareness of other parallel activities of the mind.

Whatever the skill employed in thought—that of logic, mathematics, language, spatial or musical symbols—we must not forget that it is driven by the Jamesian processes, undergoes flights and perchings, is susceptible to great variations in attention, and in general, is fueled by metaphorical and metonymic processes. It is only when the results of many parallel, fluctuating, temporal processes of perception, concept formation, memory, and attentional states are "stored" in a symbolic object—a sequence of logical propositions, a book, a work of art, a musical work—that we have the *impression* that thought is pure. Because thoughts are driven by other thoughts, by images, and by an imagined goal, we have the impression that there is a domain of *Vernunft*—a place where the thinker (in an absorbed attentional state) is nowhere and in no definable time. The path from this impression to Platonism and essentialism, both biologically untenable, is a short one.

Thought cannot be pursued except against a conscious backdrop. But a biological theory of consciousness provides only a necessary condition for thinking, not a sufficient one. Thinking is a skill woven from experience of the world, from the parallel levels and channels of perceptual and concep- tual life. In the end, it is a skill that is ultimately constrained by social and cultural values. The acquisition of this skill requires more than experience with things; it requires social, affective, and linguistic interactions. Thoughts, concepts, and beliefs are only individuated by reference to events in the outside world, and by reference to social interactions with others, particularly those involving linguistic experience.

What this means is that no amount of neuroscientific data alone can explain thinking. There is nothing mysterious or mystical about this state- ment. A neuroscientific explanation is necessary but is not sufficient as an ultimate explanation. This is comparable to the statement that, while a complete embryological account is necessary to explain how I look and act like a man, it will never explain why I *am* a man. Only an additional evolutionary account involving historical events and natural selection can provide a sufficient explanation.

At a certain practical point, therefore, attempts to reduce psychology to neuroscience must fail. Given that the pursuit of thought as a skill depends on social and cultural interaction, convention, and logic, as well as on metaphor, purely biological methods as they presently exist are insufficient. In part, this is because thought at its highest levels is recursive and symbolic. Because we are each idiosyncratic sources of semantic interpretation, and because intersubjective communication is essential for thought (whether with a real or imaginary interlocutor), we must use and study these capacities in their own right. This necessity does not, however, contradict our position that cognitive psychology cannot be properly understood without a sound, biologically based explanation of consciousness and of the processes by which meaning is embodied.

In *Acts of Meaning*, Jerome Bruner makes a strong case for the central role of the construction of meaning in human psychology. He emphasizes how the self arises from interpersonal interactions in a culture under the influence of narratives. He urges that we employ rigorous interpretive methods in social psychology. In this instance, we are our own scientific instruments, not to be replaced by measuring devices. The present work aims to provide a biological basis for the construction of meaning underlying such efforts. With this foundation, we can see how consciousness, based on evolved value systems and driven by language, leads to the extension and modification of those systems in a culture.

If we wish to investigate what *drives* thinking, however, and also what accompanies it in an individual, we must still examine that individual's biological state and explore that individual's memory at the level of thought as it prompts or motivates other thought. By its nature, this approach is a limited one. We must therefore look also to studies of intrinsic meaning, of invariant mental constructions such as mathematics, of invariance with respect to lexical substitution as found in logic, and generally to a set of rules that are socially and experientially derived. We may even leave room for philosophers without ceding to them their mistaken but time-honored privilege of applying methods of thought as the *sole* means of understanding how the mind arises.

Just as different sciences are compatible with each other but are not fully reducible to one another—one being necessary but not sufficient for the next—so a description of the matter of the mind provides a basis for the analysis of relational and symbolic matters. In making that description, one cannot help being struck by the multiple, parallel, and shifting nature of conscious states. Cognitive science's task is to find out how to interpret states concerned with symbolic modes of reasoning and states of judging or willing in which the subject is more directly aware of his or her relation to time. Even if an analysis of such matters were to succeed, however, it would not give an adequate account of the potentially limitless use of recursive modes of reasoning—induction, analogy, and formal logic. And, in any event, such an analysis would not serve to exhaust explanation in historical matters.

Biological regularities underlie all these activities. These regularities can and should be studied. But until, at some distant time, we have constructed conscious artifacts capable of speech, biological methods are too clumsy to be used to make neural correlations with the meaning of the thoughts of a "pure thinker" during a process of reasoning. We can, however, study the fundamental neural processes that underlie these acts, and we can do so without becoming property dualists. But practically speaking, it would be foolish to use only biological methods in the name of scientific purity.

Much more could be said, but it would not be illuminating for our concerns, which are to consider the biological bases of mind. It may be useful, however, to comment on a few related issues, particularly those concerned with feelings and emotions. Feelings are a part of the conscious state and are the processes that we associate with the notions of qualia as they relate to the self. They are not emotions, however, for emotions have strong cognitive components that mix feelings with willing and with judgments in an extraordinarily complicated way. Emotions may be considered the most complex of mental states or processes insofar as they mix with all other processes (usually in a very specific way, depending on the emotion). They are not made simpler by the fact that they also have historical and social bases.

What is perhaps most extraordinary about conscious human beings is their art—their ability to convey feelings and emotions symbolically and formally in external objects such as poems, paintings, or symphonies. The summaries of conscious states constrained by history, culture, specific training, and skill that are realized in works of art are not susceptible to the methods of scientific analysis. Again there is no mystification in this denial, for understanding and responding to these objects requires reference to *ourselves* in a social and symbolic mode. No external, objective analysis, even if possible, supplants the individual responses and intersubjective exchange that takes place within a given tradition and culture. A beautiful analysis of these psychological processes has been given by Suzanne Langer in her chef d'oeuvre, *Mind: An Essay on Human Feeling*.

In addition to *Vernunft* and *Verstand*, another set of German words used to characterize human knowledge makes a distinction first clearly set out by Wilhelm Dilthey. *Naturwissenschaften* refers to knowledge concerned with what used to be called the natural sciences—physics, biology, and the like. *Geisteswissenschaften* refers to fields of knowledge concerned with the social sciences, with culture, with abstract reasoning, and with studies of historical events based on symbols and feeling. In making this distinction we must not succumb to the idealism of Georg Hegel, who was among the most distinguished proponents of *Geist*. Nor must we think that psychology falls outside evolutionary biology, that there is a separate *Geist* and a separate *Natur*, for this leads to endless unnecessary complications.

James, a reflective investigator of the subjects of this chapter, had this to say about some philosophical positions on such matters:

> The whole lesson of Kantian and post-Kantian speculation is, it seems to me, the lesson of simplicity. With Kant, complication both of thought and statement was an inborn infirmity, enhanced by the musty academicism of his Königsberg existence. With Hegel it was a raging fever. Terribly, therefore, do the sour grapes which these fathers of philosophy have eaten set our teeth on edge.

Given what I have said here, I expect philosophical psychology to continue to go its own way, with this qualification: Despite the methodological differences between *Geisteswissenschaften* and *Naturwissenschaften*, psychology can no longer declare its autonomy from biology, and it must always yield to biology's findings.

I used to wonder why there were so many subjects in a university catalog. Why is knowledge so heterogeneous? The view presented here offers a possible reason. Given the parallel, constructive brain processes that underlie consciousness, given the recursive symbolic properties of language, and finally, given the irreversible historical bases for specific symbolic and artistic realizations in society and culture, there can be no fully reducible description of human knowledge. But different spheres of knowledge and different subject domains can be compatible with each other, and their bases in biological and cultural evolution can be understood. Human beings, at least in their pursuit of these different domains, seem to be doing just about what they should be doing.

A more poignant situation arises when human beings are afflicted with neural disorders. As I hope to show in the next chapter, these afflictions also reveal the enormous range of responses and the layered complexity of which the nervous system is capable.

CHAPTER 18

Diseases of the Mind:
The Reintegrated Self

A satisfactory general comprehension of neuropsychotic distur-
bances is impossible if one cannot make connections to clear
assumptions about normal mental processes.
—Sigmund Freud to Wilhelm Fliess

ental disease has always seemed mysterious. It affects the individual "soul," and as an aberration from a person's previously wit-
nessed history and behavior, it seems strange to those who know
the person if not to him- or herself. It is often difficult to trace its
causes, and while one may be convinced that it is the result of alterations
in brain function, it does not have the symptomatological "directness" of
many of the neurological disorders that also result from altered brain
function. What is the difference, and where is the dividing line? The
difference is a subtle one, but it almost always has to do with changes in
intentionality, consciousness, value, or symbolic function. A theory of
mind such as ours makes it clear that all mental diseases are based on
physical changes.

I have no intention of dealing at any length with this fascinating and
difficult subject. But to neglect all of its many facets is to lose an opportu-
nity provided by nature to check some of our models of the mind. Accordingly, I will deal here with some time-honored medical subjects from the
viewpoint taken in this book. There is no better way to reveal the multilevel controls affecting the brain and mind. First I want to discuss what

mental diseases are diseases of—whether they are physical or not. Then I want to consider why they seem different from neurological diseases. Following that, I want to take up some diseases of consciousness—both neurological and psychiatric—because they shed light on our models. Finally, I want to look at mental diseases as adaptations, as reintegrations of the self under crippling conditions. I discussed some of this material in *The Remembered Present* but not in quite the same way. The reader is invited to compare.

Freud was deeply concerned with a problem that is central to the issues with which we have been concerned. While investigating the neuroses, he strode into the thickets of intentionality. Freud thought that neuroses were those alterations of behavior or emotion that did not involve losing the ability to test reality but that did impair function or satisfaction. It did not seem to him that these disorders resulted from brain disorders. Rather, they appeared to be *functional* disorders, stemming from psychological factors, from symbolism both conscious and unconscious. They were, in his view, disorders of psychological *development.* He constructed his psychological theory based on the results of a therapeutic method, psychoanalysis. This method involves symbolic interactions between a patient and a trained analyst who has a definite theory of how the human personality is formed and how a person's ego is developed. The patient is encouraged to explore, with the analyst, mechanisms of defense and repression by using the techniques of free association, dream analysis, and the like.

Although Freud initially adopted a severe eliminative materialism (just the sort of reductionism I deplored in the last chapter), he later resorted to a kind of property dualism. While remaining a severe determinist and materialist, he held that the neuroses were nevertheless to be considered only in psychological terms. Freud's position on psychosis, in which a patient's reality testing is truly impaired and for which organic causes can be found, was more equivocal. Certain psychoses were and are considered to be "functional;" schizophrenia, some forms of manic-depressive psychosis, and paranoia, for example, have causal or etiological histories quite different from those of organic psychoses or the degenerative brain diseases that can also end in psychosis. These observations pose difficulties for any brain theory.

According to the TNGS, these difficulties arise because of the intricacies faced in sorting out the levels at which brain function is controlled. A further intricacy stems from the population nature of synaptic responses— from their diversity and individuality. The *real* problem, however, is not intricacy but the misassignment of levels of causation. All psychiatric disorders, even those traceable to difficulties in individual and social com-

179

munication, have physical causes. Given the multilevel reentrant systems of the brain that control conscious and unconscious states, it is no surprise that different causes of disease yield overlapping or similar derangement of response patterns. Sooner or later, all aberrations are reflected at synapses. But at the same time, complex signals and environmental interactions are linked to memory and behavior in patterns across all the levels discussed. It is often more useful to consider mental diseases as disorders of categorization, memory, reentry, and integration rather than as disorders of "reality testing."

One way to visualize this is to adapt the diagrams for consciousness given in chapters 11 and 12 to our present purposes (figure 18–1). If we consider, for example, that factors destroying or affecting neurons in parts of the basal ganglia lead to Parkinson's disease, or that other factors affecting the motor cortex yield a motor paralysis, we have no difficulty diagnosing the affliction as neurological—that is, as "not mental." But if the disease involves interactions among the highly parallel reentrant circuits of the brain, or alters the connections between value systems and those driving behavior (see figure 12–4), we are likely to diagnose the disorder as an alteration of mental function. In both cases, physical causes are sufficient to account for the disturbances.

The problem of mental disease may be usefully looked at in terms of alterations in reentrant pathways and in categorization. Derangement and diseases of consciousness represent rearrangements and adaptations to alterations in reentrant maps, homeostatic regions, and the cortical appendages responsible for perceptual awareness, symbolic conceptual functioning, and emotional responses.

In any event, the individual history of a person with such afflictions virtually assures that no two patients will be alike. Higher-order consciousness involves conceptual, semantic, and social integrations, all of which are involved in the construction of a social self. Many of these integrations are mediated or modulated by particular synaptic populations. It is therefore not surprising that drugs that change the functions of synapses have been found to be enormously useful in the treatment of mental disorders. But the individuality of the conscious patient results from an enormously complicated pattern of synaptic efficacies unique to him or her. The task of communicating with the patient by verbal and emotional means will therefore not be abrogated by the use of drugs alone. A combination of drugs and psychotherapy is still likely to be required in most cases.

A brain theory that views categorization, memory, and concept formation in these terms can even be useful in purely psychotherapeutic formulations. An appraisal of the TNGS by a psychiatrist formulating a theory of

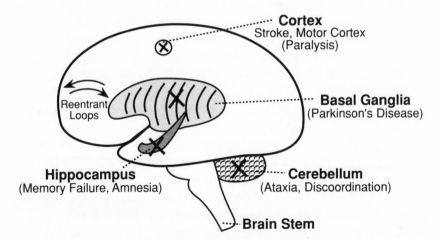

Cortex
Stroke, Motor Cortex
(Paralysis)

Reentrant
Loops

Basal Ganglia
(Parkinson's Disease)

Hippocampus
(Memory Failure, Amnesia)

Cerebellum
(Ataxia, Discoordination)

Brain Stem

NEUROLOGICAL DISEASE

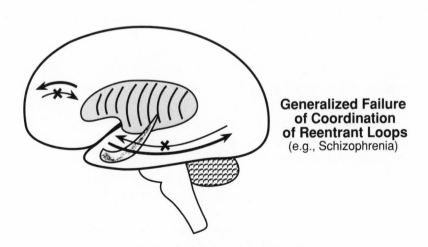

**Generalized Failure
of Coordination
of Reentrant Loops**
(e.g., Schizophrenia)

PSYCHIATRIC DISEASE
FIGURE 18–1

Diseases of the nervous system and diseases of the mind. "Neurological disease" refers to disruptions of sight, movement, and so forth, and is the result of alterations in the regions of the brain involved in these functions. "Psychiatric disease" refers to alterations in categorization, mental activity, qualia, and so forth, in which responses are symbolically deviant or in which "reality testing" is compromised. These diseases result from functional alterations at many levels, from synapses through reentrant loops. Both categories are physical in origin, and they overlap. Psychiatric diseases affect categorization, memory, and symbolic processes more extensively via reentrant loops.

psychoanalytic treatment was published several years ago. In *Other Times, Other Realities*, Arnold Modell employs the idea of memory as recategorization to reevaluate the nature of the transactions between the patient and therapist. He revives the term *Nachträglichkeit*, which Freud applied to the idea that a memory is retranscribed as a result of subsequent experience. Modell points out that the ego is a structure engaged in the processing and reorganizing of time, which he links to the idea of memory as recategorization (see chapters 10 and 16).

In his critique of concepts related to the transference relation in the psychoanalytic setting, Modell proposes that a selectionist view of brain function offers an alternative interpretation of the repetition compulsion described by Freud. Modell proposes that the recreation of a categorical memory is a fundamental biological principle, one that under certain circumstances supersedes Freud's pleasure principle. He suggests that "the compulsion to repeat represents a compulsion to seek a perceptual identity between present and past objects." In doing so, the patient is aware to varying degrees of the relation of the self to time, a subject touched on in chapter 16. Modell also points out that, in the psychoanalytic process, thinking in metaphor is the currency of the mind. Finally, he suggests that the view of memory proposed in the TNGS, which replaces Freud's view of fixed memory traces, opens up a new way of looking at the treatment setting. In this view, the treatment setting is designed "to accentuate multiple levels of reality, which in turn enlarges the potential for both old perceptions and retranscriptions of new perceptions."

Another psychiatrist, Edward Hundert, has also used the TNGS as a touchstone for relating psychiatry, philosophy, and neuroscience. He has proposed that all three disciplines must be related through such a theory to assure that individuals are viewed in a sufficiently broad way. Interested readers are encouraged to compare his remarks with the views I presented in the previous chapter and throughout this book.

Although I am hardly qualified to speak with authority about matters of psychoanalytic theory, it seems to me to be important that attempts are being made to relate psychoanalytic theory to a physically based brain theory concerned with problems of categorization. Attempts to relate pharmacological effects to brain theories are somewhat more prevalent. We need both kinds of effort if we are to have a psychiatry that is solidly based in biology.

With these general comments, I will turn to a specific set of neurological diseases and one psychiatric disease. I do so because both kinds of disease shed light on the proposals I have made about the importance of reentrant circuits to consciousness. The neurological disorders I will discuss are a set of syndromes that involve what has been called implicit memory. The psychiatric disease is schizophrenia, the most florid, varied, and mystifying of psychoses.

First, the neuropsychological syndromes (here the reader will have to bear with some clinical jargon; I will translate as I go along). Daniel Schacter and his colleagues have studied a set of dissociations between explicit, conscious access to knowledge and the implicit ability to perform a task. This kind of dissociation—near-normal implicit knowledge with severely impaired explicit knowledge—has been observed in patients with a variety of diseases. Before listing them and describing some cases, it might be best to give a concrete example: blindsight. Patients with blindsight distinguish visual stimuli in space even though they are blind in the part of the visual field in which these stimuli are presented. These patients perform tasks of perceptual discrimination even though they are perceptually unaware or unconscious of being able to do so.

Such dissociations have been observed in patients with amnesia, dyslexia (the inability to read certain texts), aphasia (the inability to express or produce intelligible speech), prosopagnosia (the inability to recognize faces), hemineglect (the inability to attend to the egocentric side of space opposite to a damaged cerebral hemisphere), and anosognosia (the unawareness of or denial of gross neurological defects even when presented with direct evidence for their existence).

This is an extraordinary list. Amnesiacs show learning responses to priming cues during performance that indicate the presence of specific knowledge even though the patients deny awareness of such knowledge. Prosopagnosics show implicit knowledge of faces that one knows they have seen before the onset of their disease, but of which they now have no conscious awareness. Under certain circumstances, dyslexics read texts without the awareness that they can do so. And in patients with hemineglect, information presented on the neglected side affects the patient's performance of a task without his or her conscious knowledge.

The lesions that underlie these disorders are all different. There is no global change of consciousness in any of these cases. Indeed, the patients behave quite normally outside the domains exhibiting the defect; in other words, in each case the defect is domain specific. There does not seem to be any evidence that language impairments are responsible for the patients' dissociative responses. Above all, there is no evidence that patients with these syndromes suffer from neurosis or psychosis.

These diseases are diseases of consciousness. They can be explained by assuming that what has been affected is the special reentrant loop connecting a value-category memory to classification couples carrying out perceptual categorization (figure 18–1; see also figure 12–4). Notice that this interference will not, in general, affect *other* pathways connecting with value-category memory and mediating the performance of a specific task. The result is a domain-specific deletion from the conscious "scene" but *not* from the repertoire of the individuals' capabilities to perform the task. If these reentrant pathways were clipped one by one across all the modalities, one wonders whether any primary awareness would remain. This is unlikely to occur except as a result of massive injury, and on ethical grounds the experiment could hardly be performed even if some less traumatic means of bringing it about were available.

I have described these *neurological* disturbances to show that they, like mental diseases proper, can also be *mental*—that is, they can affect intentionality. To illustrate these disturbances even more dramatically, I end this chapter by describing a case of dissociation that is downright mysterious, that of anosognosia. But first, let us turn to the most flagrant, polymorphic, and mysterious of psychoses: schizophrenia. Schizophrenia is characterized by mixtures of bizarre symptoms. These include third-person auditory hallucinations, delusions of control by alien forces, ideas of influence, ideas of reference, thought echo, thought broadcast, thought block, and thought withdrawal. In the acute stages of this syndrome, the symptoms are accompanied by a dreamlike state and a slight clouding of consciousness as well as by perplexity. Patients experience a barrage of signals that make sense only in fragments and across islands of awareness. Some patients show blunting of affect. Some are easily distracted, misjudge perceptual signals, have visual illusions, or are not able to discriminate gestalt figures. Others show poor judgment, are slow to respond, or perseverate. In extreme forms of the disease, patients are sometimes catatonic; they are unresponsive while maintaining bizarre postures for long periods of time. On recovery from acute episodes, some patients remember what was said or done by others during the catatonic state, while others have a poor memory of the acute experience.

Schizophrenia is a diverse and protean disease of consciousness, affecting perception, thought, and qualia. It is a moving and sorrowful experience to see a patient, with his or her unique constellation of symptoms, in the grip of this affliction. It has not been easy to account for the diversity of symptoms, the individual characteristics, and the bizarre features of this disease.

I have suggested the possibility that schizophrenia is a generalized disease of reentry (see figure 18–1, bottom). I have hypothesized that a disorder in the production of or response to several neurotransmitters could cause a general disabling of communications between reentrant maps. If there is a failure in appropriate mapping or an asynchrony between maps, imaging may predominate over perceptual input, or different modalities may no longer be coordinated. This could lead to hallucinations and failure to coordinate real-world signals. The same disorder in another patient may lead to disturbances in the reentrant engagement between different conceptual systems and the organs of succession. In yet others, there may be disorders in reentrant linkages between areas involved in the lexicon, conceptual centers, and those that mediate imagery.

Historical and individual factors involving repertoire variation explain the different responses found in different schizophrenic patients. Each patient has a unique pattern and responds differently to sites of reentrant disorder. I am not saying that all neurons are normal in schizophrenia (I doubt that this is so) but only that the *main* psychological defect is the result of a defect in reentrant mapping. This may be caused by any of the factors that alter individual maps or their connections, including neuronal disorder or loss.

It is not difficult to see how a patient who still has higher-order consciousness would, if afflicted with this disorder, try to adapt as a self to what he or she perceives. This behavior would obviously appear abnormal by normative and physiological criteria, but it is likely that the patient's overall response is still an attempt at adaptation, at reintegration. That it is not the best adaptation or that it may be destructive to the self and others is not in question. But the mind of a schizophrenic, viewed from the extended TNGS, does make sense, particularly if one knows the history of the individual patient. Alas, the "sense" is made in terms of the predictions of the theory, not in the social or affective terms that are specific to a given society.

We underestimate the faculties of psychotics as readily as we misread the apparent eccentricities of normal people. I do not know where I heard or read of the man in Paris during the Nazi occupation, who, knowing that the Gestapo was closing in on him, decided that the last place they would search for him would be in a hospital for the insane. So he cultivated the art of behaving crazily, and after a "hallucinatory" episode in the street he was committed. Life went on without terror for quite a while, and from time to time he displayed deliberately bizarre behavior in front of the doctors and his fellow patients. One day two sinister men in long black leather coats appeared at the door of his room, accompanied by the chief superintendent. Certain that the Gestapo had arrived, he leaped up, assumed a bizarre posture, rolled his eyes, and began to emit strange yelping noises. Whereupon the man in the next bed, who spent most of his time in a trance-like state, opened his eyes and said to him firmly, "Taisez-vous, simulateur" ("Shut up, faker").

As for normal "simulation" (but not real faking), I recall my dear friend, the late Lars Onsager, one of the most extraordinary physical chemists of our time, indeed the most phenomenally gifted scientist I have ever known. Within five minutes of taking his seat at some arcane lecture, his head would loll to one side in a posture of sleep. But if the lecturer erred in an equation, Lars was likely to get up, stroll to the board, erase the error, correct it, smile, go back to his chair, and fall asleep again. I once asked him, "Lars, when people ask you deep questions, why do you grin and giggle and nod your head and say incomprehensible things?" He became grave, almost stern, and said, "I'm lazy." "Lazy!" I exclaimed, "I don't get it." To which he responded, "I want to answer *my* questions, not *their* questions."

I have deliberately discussed two extreme examples from the "neurological camp" and the "psychiatric camp"—implicit–explicit dissociation and schizophrenia. Both are examples of diseases of consciousness. We must beware of Cartesianism in analyzing patients with either type of disease. To drive that point home, I will describe a case reported by the Italian neurologist Eduardo Bisiach. This case presents as significant a challenge to brain theory as the one that faced Freud in the early days of psychoanalysis when he confronted his patient Anna O.

Bisiach's patient had anosognosia; that is, he denied the existence of a neurological defect even when presented with direct evidence for its existence. In this case, the syndrome accompanied left hemiplegia (paralysis on the left side) and left hemianopia (inability to see the left visual field). The cause was a sudden vascular episode affecting his right temporal, parietal, and occipital brain areas. (Roughly speaking, these cortical areas mediate visual tasks as well as some motor capabilities.)

The patient was intelligent, responsive, and in no obvious way emotionally upset. He showed no evidence of a speech disorder. But he was anosognosic to his deficiencies both of sight and of movement. When questioned about left-sided tasks that he did not actually perform, the patient claimed to have performed them. (His left-sided paralysis obviously made their performance impossible.) The patient's paralyzed left hand was then placed in the hands of his examiner and positioned in the patient's right visual field so that he could see it. He was then asked whose hand it was. He claimed it was the examiner's. When asked about the three-handed discrepancy, the patient's response revealed flawless logic: "A hand is an extremity of an arm. Since you have three arms, it follows that you have three hands."

If we disqualify suggestions that this patient was neurotic, psychotic, or had a language disorder, we are forced to admit to an extraordinary conclusion: This is that consciousness based on language can be altered by the removal of brain sources of nonverbal signals in the individual. What is striking is that this patient reintegrated his entire semantic interpretation of reality without emotional disturbance. He underwent a radical *conceptual* rearrangement and reintegration. One would have thought that if his body image and categorical capability were "stored" in memory, the contrast of that stored memory with his present perceptual and motor state would have led either to a consistent "realistic" report of his actual plight or to great conflict and emotional disturbance. But if our theory is correct, the patient has no such fixed memory. He cannot attend to parts of personal space, and he has undergone a conceptual and semantic reintegration that not only reflects this incapacity but in some sense builds an adaptive picture around it.

This case clearly indicates how changes in intentionality can accompany neurological disease. It also serves as a deep challenge to our notions of mind. How is the self reintegrated? In this connection, I should also mention the remarkable responses of a split-brained individual, the left brain of whom appears to belong to a more or less normal person with higher-order consciousness. In some cases, the right brain responds to visual words with left-handed responses spelled out in Scrabble letters. Sometimes the responses are concordant with those of the left brain, sometimes not. One must at least consider the possibility that the right brain is mediating primary consciousness. Unfortunately, a complete analysis is blurred by such factors as past learning and the previous connectivity of the hemispheres. In any case, the *person* reporting with higher-order consciousness is seated in the left hemisphere and interprets all events adaptively.

These observations pose problems closely related to those studied by Freud. What governs the disposition of conscious and unconscious responses so as to allow integration into personhood? What is the minimum apparatus required for the appearance of higher-order consciousness? What governs the course of reintegration when the individual is affected by disease?

The study of mental disease at all levels is obviously as important to an understanding of how the brain works as it is to an understanding of what it means to be an individual in a society. Given its cultural significance to us, the study of mental disease has obvious practical significance. However, the complexities of these studies are enormous, and it is unlikely that the normal and abnormal workings of the brain can be unraveled by psychiatry alone. Many disciplines are needed. At the furthest reach from psychiatry, attempts have been made to synthesize objects and artifacts that have psychological functions and manifest intentionality. If these efforts are successful, they will play an important part in helping us understand our place in nature, both in health and in disease. So I turn now from the study of intentional humans to the possibility of creating intentional things. This is an exciting prospect and, whatever its limitations, one of great practical and theoretical importance.

CHAPTER 19

Is It Possible to Construct a Conscious Artifact?

It is clear that there is but one substance in this world, and that man is its ultimate expression. Compared to monkeys and the cleverest of animals he is just as Huygen's planet clock is to a watch of King Julien. If more wheels and springs are needed to show the motion of the planets than are required for showing and repeating the hours; and if Vaucanson needed more artistry in producing a flautist than a duck, his art would have been even harder put to produce a 'talker', and such a machine, especially in the hands of this new kind of Prometheus, must no longer be thought of as impossible.

—Julien Offray de la Mettrie

After the discussion of human issues in the last several chapters, the title of this one may seem incongruous. My purpose is to consider whether a knowledge of brain function will allow us to construct intentional objects. I also wish to raise the possibility that the only way we may be able to integrate our knowledge of the brain effectively, given all its levels, is by synthesizing artifacts. To do this, we need the most advanced kind of computers. The construction of "conscious" artifacts has a meager but definite history, one to which the epigraph and figure 19–1 bear testimony.

Indeed, is it not contradictory to suggest a need for computers in a book maintaining that the brain is *not* a computer? To get at the answers to this question, I have to say a bit about computers. Then I want to consider

FIGURE 19–1

Jacques de Vaucanson (1709–1782), a famous constructor of artifacts imitating behavior, shown with his duck. The construction quacked, waddled, and had "intestinal function."

whether it is possible to construct several kinds of artifacts: a perception machine, an artifact with primary consciousness, and one with higher-order consciousness. If the answer is yes to any of these questions, there *are* moral issues to be considered, as is the case in the application of any scientific finding.

Computers are logic engines; in principle, they can carry out any effective procedure that is specified unequivocally by a set of instructions and produces a unique result for a given problem. I have said that the brain is not a computer and that the world is not so unequivocally specified that it could act as a set of instructions. Yet computers can be used to *simulate* parts of brains and even to help build perception machines based on selection rather than instruction.

I can resolve the apparent contradiction by pointing out what a simulation does. In a simulation, a program is written that specifies the required structural properties and operating principles of the entity to be simulated. The program is so constructed that when it is run, parts of the entity that is simulated in its entirety will carry out their proper functions. If, for example, I want to simulate a 747 airplane flying into turbulent conditions, I have to put into the program the design features of the airplane as well as the principles that allow it to fly—the properties of the airfoils, the power characteristics of lift for a certain weight, and so on. If the program is well designed, I can "fly" the aircraft under both smooth and turbulent conditions. My goal might be to see whether the plane loses control under certain conditions, or whether a wing vibrates too much and shears off. If successful, the whole exercise is cheaper and more informative than running a real-world model of the plane in a wind tunnel.

Can a selectional system be simulated? The answer must be split into two parts. If I take a *particular* animal that is the result of evolutionary and developmental selection, so that I already know its structure and the principles governing its selective processes, I can simulate the animal's structure in a computer. But a system undergoing selection has two parts: the animal or organ, and the environment or world (see chapter 8). No instructions come from events of the world to the system on which selection occurs. Moreover, events occurring in an environment or a world are unpredictable. How then do we simulate events and their effects on selection? One way is as follows:

1. Simulate the organ or the animal as described above, making provision for the fact that, as a selective system, it contains a generator of diversity—mutations, alterations in neural wiring, or synaptic changes that are unpredictable.
2. Independently simulate a world or environment constrained by known physical principles, but allow for the occurrence of unpredictable events.
3. Let the simulated organ or animal interact with the simulated world or the real world without prior information transfer, so that selection can take place.
4. See what happens.

Now we have a situation in which unpredictable variation is occurring in each of two separate systems or domains. Moreover, if we allow these domains to interact in the simulation, there is no way of predicting the outcome beforehand (other than by performing the same simulation in another computer). Therefore, no effective procedure simpler than the simulation itself can predict the outcome. Put another way, the program used for each domain *does* specify constraints and structures, but it *cannot* specify the exact way in which these will act under conditions of variation and selection. Variational conditions are placed in the simulation by a technique called a pseudo-random number generator. This is a formula that produces numbers in a way that simulates randomness! It doesn't succeed absolutely; if we wanted to capture randomness absolutely, we could hook up a radioactive source emitting alpha particles, for example, to a counter that would *then* be hooked up to the computer. But it appears likely that if we choose two separate random number generators, one for variation in the animal and one for variation in the environment, and allow selection to occur, we are likely to avoid introducing a predictable bias into the system. After all, the two systems have no way of "knowing" what variation will match what, and we can always keep changing each random number generator.

Under these circumstances, we cannot specify an effective procedure *for the consequences of selection* that is independent of our choice of pseudo-random number generators. Insofar as this is so, it is not meaningful to describe the system and its future results *as a whole* as a computer (or Turing machine). Any particular past result of selection *once we know it* can of course be so specified. By the procedure described above we have added features to our simulating computer that convert its performance into something that is not strictly that of a computer. This is the result of choosing methods for generating random numbers that do not couple in predictable ways to the sequences of events in the systems being simulated, so that the theoretical probability of a future event lies in the sequence of pseudo-random number generators and not in the simulation itself.

Now we are in a position to tackle our first question: Is it possible to construct a perception machine? Yes, although the ones that have been constructed so far are primitive. I have already presented some data on the performance of one the first of these in chapter 9. This artifact, Darwin III, has a four-jointed arm with touch receptors on the part of its arm distal to the last joint, kinesthetic neurons in its joints, and a movable eye. It contains simulated neurons in numerous repertoires that show diversity in both local connections and synaptic strengths. Although it sits still, it can move its eye and arm in any pattern possible within the bounds imposed by its mechanical arrangement. Objects in a world of randomly chosen shapes move at random past its field of vision and occasionally within reach of its arm and touch. The synaptic strengths of its neurons are initially set by a random number generator. After encountering objects (and responding to them), it displays behavior that appears very much like perceptual categorization (see figure 9–6). This is the case as long as its neural circuits have been constructed to respond to value (for example, light is better than darkness, or touch is better than no touch). Thus, it carries out categorization on value.

A word about this is in order—the basis for value *is* programmed into the design of the machine. But this value is *not* the same as any category, which is not so programmed. The programming of value is allowed because value is considered to have resulted from the *evolutionary* selection of preferences of a particular type because they conferred selective advantages on the individuals of a species. If we were to simulate a cat, we might put in value systems or circuit constraints that made movements leading to fur licking (as detected by simple parts of the cat's own nervous system) more likely to be rewarded. The effects of such constraints on behavior would ultimately have to be observed, not programmed. In our automata, we have *not* programmed in the kinds of categorization that result from actual somatic selection, because these changes are epigenetic.

There is nothing, by the way, especially mysterious about value. For instance, to give Darwin III values positive for lighted objects in its central vision, specialized neurons are constructed with inputs more densely wired in the central part of its "retina" and less densely in the periphery. The image of a lighted object falling on the central retina produces strong responses in these neurons. These responses are then transmitted to the vicinity of the synapses that connect visual neurons with the motor neurons that move the eye. Activity in these systems leaves a "chemical trace." While these traces last and following any movement that brings the stimulus toward the central region of the eye, a value signal leads probablistically to the strengthening of the synapses involved. This increases the likelihood that similar motions will take place the next time a stimulus appears in a similar position.

The behavior of Darwin III is quite limited. It does not categorize across a broad set of stimulus characteristics, and it does not show true learning, although experiments providing it with a "taste" system have suggested that it modifies its selection patterns for objects after the alteration of its values. In any case, to test such behaviors, one wants to have a much richer environment, one composed of real-world objects. A perception machine with its nervous system simulated in a supercomputer and with its "real" eyes and motor apparatus in a "creature" "living" in a different room is currently under construction (figure 19–2). This artifact, NOMAD (Neurally Organized Multiply Adaptive Device), is in touch by television and radio with its brain, a brain more complex than that of Darwin III but still designed as a selectional system. The tasks set for NOMAD couple categorization to learning—picking out variously shaped objects it considers to have value, for example. Unlike what happens in ordinary robots, these choices are not programmed; they are the result of training.

I have called the study of such devices *noetics* from the Greek *noein*, to perceive. Unlike cybernetic devices that are able to adapt within fixed design constraints, and unlike robotic devices that use cybernetic principles under programmed control, noetic devices act on their environment by selectional means and by categorization on value. This field is in its early stages, but already it promises to teach us much about how to study the layers and loops of neurally organized systems. In time, noetics will also have considerable practical importance.

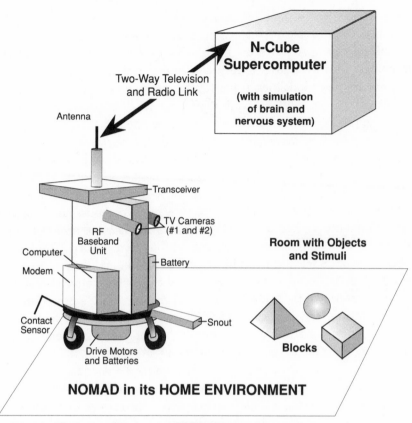

FIGURE 19–2

NOMAD (Neurally Organized Multiply Adaptive Device) is a real-world artifact constructed on principles similar to those of Darwin III. Although NOMAD's brain is simulated on a powerful supercomputer, this "brain" does not act like a computer. NOMAD "lives" at the Neurosciences Institute and is the first nonliving thing capable of "learning" in the biological sense of the word. With its "snout" it picks up magnetic blocks of different shapes and colors that yield "value" (electric stimulation) on contact. While NOMAD looks like a robot, it does not operate like a robot under the strict control of a program. It operates like a noetic device, one that is neurally organized and works according to selectionist principles. The neural impulses transmitted to NOMAD from its simulated brain (impulses that in an animal would activate muscles) are translated instead into signals for NOMAD's wheels by an on-board computer.

Can we extend these notions to the construction of a primarily conscious artifact? The answer is not straightforward. But one might hazard a guess and say, in all likelihood, yes. In principle there is no reason why one could not by selective principles simulate a brain that has primary consciousness, provided that the stimulation has the appropriate parts. But there is much to be done before a conscious artifact can be successfully designed. For example, no one has yet been able to simulate a brain system capable of concepts and thus of the *reconstruction* of portions of global mappings. This in itself is a very challenging task. Add that one needs multiple sensory modalities, sophisticated motor appendages, and a lot of simulated neurons, and it is not at all clear whether presently available supercomputers and their memories are up to the task.

If the proposed model for primary consciousness is correct, such a simulation, possible in principle, can be used to test the self-consistency of these ideas. How? By enabling us to see whether an artifact capable of correlating a scene by reentry between value-category memory and perceptual categorizations behaves in such a way as to choose combinations of causally unconnected outside events *for its own adaptive needs*—according to its own assignment of salience and its own history. The test of efficacy involves leaving the artifact's circuits intact and then cutting key reentrant loops, one at a time, to see what, if any, deteriorative effects such a disruption has on the artifact's adaptive behavior. (Such a procedure is a bit like testing the implicit–explicit dissociations discussed in the last chapter.) This illustrates one of the main values of simulations, one already proven in the case of perception machines like Darwin III. Given the complexity of neural patterns and behavior and their many levels of interaction, only a fast computer with a huge memory storage could hold all the patterns necessary at each level for scientific review. Computers are not appropriate models of brains, but they are the most powerful heuristic tools we have with which to try to understand the matter of the mind.

Given what I have just said, you have probably guessed the answer to the last of our questions concerning artifacts with higher-order consciousness. It may be possible to construct such artifacts someday, but right now it is so unlikely as to be unworthy of too much reflection. Not only will an artifact with primary consciousness already have to have been made, but we will also have to have understood how *at least two* such systems could intend something "to be what it is for each other under the auspices of a symbol," as the writer Walker Percy put it (see the Postscript). In other words, artifacts with higher-order consciousness would have to have language and the equivalent of behavior in a speech community. A great deal still remains to be understood about the organization of linguistic memories, and a quick solution to this problem does not seem likely. For now, we can relax in the knowledge that, so far, we remain the only known systems with linguistically based higher-order consciousness, and competing artifacts remain a long way off.

In principle, however, there is no reason to believe that we will not be able to construct such artifacts someday. Whether we should or not is another matter. The moral issues are fraught with difficult choices and unpredictable consequences. We have enough to concern ourselves with in the human environment to justify suspension of judgment and thought on the matter of conscious artifacts for a bit. There are more urgent tasks at hand.

In thinking about these matters, we must remember how young a truly integrated science of the mind is. Of course, observational psychology is one of the oldest of "sciences." But psychologically sophisticated neurobiology is in its infancy. So we may have to wait a while, indeed perhaps a long while, for the kinds of developments I have discussed in this chapter. My personal belief is that the construction of conscious artifacts will take place, under enlightened circumstances and with due concern for human welfare. But it will take a long time. The pop artist Andy Warhol once approached me at a party and told me that he collected scientific journals, but couldn't understand them. He drifted away, then came back and said, "Do you mind if I ask you a question?" "Of course not," I replied. He asked, "Why does science take so long?" I said, "Mr. Warhol, when you do a picture of Marilyn Monroe, does it have to be exactly like her, as close to being her as you can make it?" He said, "Oh no. And anyhow, I have this place called the Factory where my helpers do it." I said, "Well, in science it has to be exact, as exact as you can make it." He looked at me with limp sympathy and said, "Isn't that terrible?"

One of the issues I have left out of the discussion so far is what some philosophers have called "chauvinism" versus "liberalism." *Must* artifacts of the kind I have described be made of organic molecules? For perception machines, the answer is already in hand: no. But the close imitation of uniquely biological structures *will* be required. If our position on the mind is correct, however, liberalism will never be the order of the day. Even with a complete knowledge of brain structures, the bet is that we will not be able to design software for consciousness to run successfully on *any* sufficiently powerful computer in the manner demanded by functionalism (see the Postscript). The constraints of morphology and selectionism run counter to these hopes.

So the answer to the question posed in the title of this chapter is: in principle, yes, but the practical problems involved in "making" higher-order consciousness are so far out of reach that we needn't concern ourselves with them right now. As for the idea of a primarily conscious artifact, a somewhat stronger yes but with the caveat that much remains to be learned about how a neural system mediates concepts in a body. As for artifacts without consciousness and capable of categorizing, prototypical perception "machines" already exist.

We have come a long way with computers in less than fifty years by imitating just *one* brain function: logic. There is no reason that we should fail in the attempt to imitate other brain functions within the next decade or so. Given the promise of research on synthetic neural modeling of the nervous system (of the kind used in Darwin III), we may soon be able to consider what kind of performance would emerge if we hooked ten perception (or P) machines capable of categorizing novelty to a Turing (or T) machine capable of logic. The combination, a PT machine, may behave with respect to the recognition of novelty roughly like a hunter and his dogs, provided that the P machines are well trained and the T machine is well programmed by a human operator. The results from computers hooked to NOMADs or noetic devices will, if successful, have enormous practical and social implications. I do not know how close to realization this kind of thing is, but I do know, as is usual in science, that we are in for some surprises.

CHAPTER 20

Symmetry and Memory: On the Ultimate Origins of Mind

The most incomprehensible thing about the universe is that it is comprehensible.

—Albert Einstein

'L'Homme pense; donc je suis,' dit l' Univers.

—Paul Valéry

osmology has formed part of the myth and the science of many civilizations. In it the mind has always played a central role, whether interior, exterior, or ulterior. It is natural for creatures like ourselves to wonder how everything came about, how we ourselves got here, and how we could come to be aware of the world in which we find ourselves.

The religious cosmologies of the past have been replaced in some cultures by a scientific cosmology, one with remarkable ties to the farthest reaches of theoretical physics. But as grand and mysterious and beautiful as this scientific cosmology is, it has no inherent principle that would lead us to ourselves: observers who are conscious, who formulate physics and relate it to cosmology, and who have the urge to place ourselves within the scientific world view we have constructed. Even a "theory of everything," as some physicists call it, would be incomplete if it did not provide us with such a principle.

In this book I have maintained that mind has arisen in a very definite way through the workings of evolutionary morphology. I have attempted to show that consciousness has arisen, at least in this little speck of the cosmos, at a particular historical time. That it emerges from definite material arrangements in the brain does not mean that it is identical to them, for, as we have seen, consciousness depends on relations with the environment and, in its highest order, on symbols and language in a society.

Higher-order consciousness leads to a rich cognitive, affective, and imaginative domain—feelings (qualia), thought, emotions, self-awareness, will, and imagination. It can construct artificial mental objects such as fantasies. In culture, it leads to studies of the stable relations among events (science) and among mental objects (mathematics), as well as to studies of the relations among sentences that refer to events and mental objects (logic).

The way I have proposed that the mind arose in nature may seem strange. This is partly because it does not seem to have come about in the same way that our most cherished constructions and inventions have arisen—through orderly relations such as those guiding the logic, arithmetic, and physics that have led us to build computers and other information-based devices.

This does not mean, however, that a deep principle in nature underlying the evolution of consciousness cannot be found. In this final chapter, I want to speculate on what that principle might be and connect it to a more firmly grounded principle that most physicists would agree is among the most fundamental in physics and cosmology. Then I want to ask how the two principles together might govern future scientific thinking and our view of how we fit into the cosmos.

Physics and biology will "correspond" with each other in an intimate way, certainly in the next century and possibly even sooner than that. At the very least, they will have an exchange about how the human observer influences physical measurements and how the observers' perceptions relate to their physical descriptions. This is a key problem in quantum mechanics (see the Postscript).

These predictive remarks may sound vague and utopian, but readers will have to make their own judgments after they think about what has been and will be said here. As is the case for every scientific specialist, my knowledge and experience have sometimes been less than satisfactory in trying to judge the relations between various fields of knowledge. Indeed, comparisons of specialized experience among scientists sometimes leads to impasse. I was once asked by the man who taught me quantum mechanics and statistical mechanics, George Uhlenbeck, a very great physicist, to introduce him to an equally great biochemist, Fritz Lipmann, who tutored me in graduate school. I arranged a dinner for the three of us. Fritz was eating his soup with gusto, when George, a man of serious dignity, said, "I have concluded from a calculation of Gibbs potentials in various phases that life happened only once in the entire history of the universe." Fritz kept pushing spoonfuls of soup into his mouth, but between two of them, he looked up and said emphatically in his charming accent, "Happens all ze time." George remonstrated with physical arguments of great sophistication. Fritz listened, kept eating, and finally emptied his bowl. He put his spoon down and said: "Ven ve ground up ze pigeon breast muscle, zey said ve vouldn't get ze oxidative phosphorylation. Ven ve got ze microsome, zey said ve couldn't get ze protein synthesis. Ve did. Don't vorry, it happens all ze time." They smiled at each other but I am sure that each left the table frustrated with the intellectual limits of the other's specialty.

This past century has seen perhaps the greatest revolution in scientific thought of all time. The revolution is not just in the adoption of unusual concepts that are necessary to understand the universe and its fundamental particles. It is in our notion of the scientific observer and in the extension of the generality of scientific thinking. Beginning with Planck's discovery of the quantum, and continuing through the early formulations of quantum theory by Bohr and Heisenberg and of the theory of relativity by Einstein, observers were no longer seen as fully detached from their measurements. In quantum measurements, the way in which observers choose to arrange their apparatus determines the outcome. In relativity theory, observers' measurements of time and length depend on their relative velocity and acceleration. Thus, the observers' conscious choices in the one case and their physical location in the other must explicitly be taken into account. The outcome of efforts to rationalize these findings is well known: Quantum mechanics and the general theory of relativity are the two grandest theoretical constructions in science. They range in their descriptions from the smallest and most short-lived fundamental particles to the edge of the measurable universe.

What is perhaps not appreciated outside the community of physics is that underlying both of these descriptions is a key mathematical principle: symmetry. This is not the place to go into the mathematics, but I shall try to say enough here to give you a glimpse of it. Symmetry is a stunning example of how a rationally derived mathematical argument can be applied to descriptions of nature and lead to insights of the greatest generality. I want to discuss symmetry a bit because I plan to contrast it with another principle I believe underlies the mind, and indeed all of biology, the principle of memory. Later I will argue that an understanding of these two principles, interacting in a tense harmony, will allow us to see more clearly the place of our minds in nature.

We are all familiar with symmetry from daily experience. As creatures, we are roughly bilaterally symmetrical. We know that our mirror images have certain properties and that our right and left hands are mirror images of each other (figure 20–1). No operation in the real world will convert a right hand into a left without destruction. But we can convert a right-handed glove into a left-handed one if we turn it inside out. This suggests that certain *operations* are necessary to reveal certain types of symmetry.

Symmetry principles and the rules governing these operations are embedded in the mathematical theory of groups, which plays an essential role in the construction of advanced physical theories. This mathematical theory was formulated in the early nineteenth century by a young French genius, Evariste Galois, whose life was cut short at age twenty and a half in a duel over a woman.

Galois's ideas about groups revealed the general insolubility of quintic equations (polynomial equations of the fifth degree; most nonspecialists will have stopped at quadratics, degree two). But it turns out that his ideas also have the most general applicability. The group of mirror reflections is concerned with discontinuous change, as we have seen (figure 20–1). Other groups deal with continuous symmetries—for example, translations in space. (This theory was advanced spectacularly in the second half of the nineteenth century by the Norwegian mathematician Sophus Lie.) The highest symmetries are generally possessed by relatively featureless objects such as a circle in two-dimensional space and a sphere in three-dimensional space. In general (but not always), the addition of features to such objects results in a lower symmetry.

Here we come to one of the formal constraints that bring deep insight to the laws of physics. This is the connection between the ideas of symmetry and the so-called conservation laws of physics. The study of physics has revealed that a number of fundamental quantities are conserved in mechanics and in both electrical and particle fields. Mass–energy, momentum, and spin are each governed by conservation laws requiring that each is neither created nor destroyed within the whole context of a physical description. Electric charge follows a conservation law: The number of positively charged particles in the universe is equal to the number of negatively charged particles. This law has analogies among fundamental particles. The number that counts protons and other particles is conserved, as is the number that counts electrons and related particles.

The consequences of applying the principle of symmetry are truly beautiful, for the different laws set limits on the ways in which these

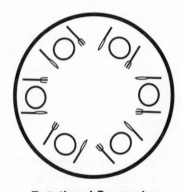

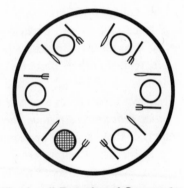

Rotational Symmetry
(Six turns bring one back
to the original place setting)

"Broken" Rotational Symmetry
(The switch of a knife and a fork
breaks the symmetry)

Axis

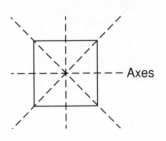

Mirror Symmetry

FIGURE 20–1
Some kinds of symmetry.

201

particles will interact with each other. In other words, the rules describing the *interactions* between particles are constrained by conservation principles. As a result, in some cases particles can only be created or destroyed in pairs, while in others particles can be created or destroyed without these constraints.

Thus, we arrive at one of the grand themes of physics: *There is a deep connection between conservation laws and symmetry.* Empty space and time are symmetric; that is, they appear the same under many kinds of change. Space is the same regardless of translations, rotations, and changes in direction. When reversed, time is the same in either direction (I mean the time of the physicists, not your own personal sense of time). Indeed, in quantum mechanics and relativity theory, the *laws* of motion themselves are invariant under symmetry operations such as rotation and translation. These invariances assure that the outcomes of physical events do not depend on the systems of coordinates used to measure them.

Without the application of a force, a body or a particle will not alter its velocity and direction of motion (its momentum) or its energy. The German mathematician Emmy Noether first showed that the conservation of these quantities can be *formally* identified with symmetry principles. For example, the conservation of momentum corresponds to the symmetry of space under translation. The conservation of angular momentum corresponds to the symmetry of space under rotation. The conservation of energy corresponds to the symmetry of time under reversal of direction. (Time reversal cannot actually be carried out, but the physical laws can be checked for their invariance under such operations.)

It was Einstein who first understood the significance of the invariance of the laws of physics and, thus, of their symmetry. Indeed, his general theory of relativity may be considered a means by which to search for conditions of *absolute* invariance!

More recently, a series of discoveries has made it possible to envision unifying all particle interactions under a single theory, a grand unification theory (or GUT). This has not yet been accomplished for all four forces of nature—the strong, weak, electromagnetic, and gravitational forces. But partial theories have been posited that are stunning, theories not even envisionable twenty years ago. If there is a major language of these theories, it is the language of symmetry. The hope is that eventually the whole of nature (read "physics") will be described by a symmetry that leads to all fields and forces in a unique manner.

It would take us too far afield to discuss all these matters. I will, however, mention two notions that are fundamental to physicists' efforts to construct a unified field theory. These notions—also essential to modern cosmology—are the ideas of local gauge symmetry and spontaneous symmetry breaking. Local symmetry can be contrasted with global symmetry. To leave a global symmetry invariant in a given domain, any transformation that takes place must occur *everywhere*. Local symmetry, by contrast, allows different transformations to occur in different parts of space and time. A theory about local symmetry developed by C. N. Yang and F. E. Mills played a key role in the success of later unification attempts. For example, if one considers a field and if one wants to achieve invariance under a change of local symmetry, things must be arranged so that *another* field will act to compensate exactly for any local changes introduced by the first operation.

To understand the gist of symmetry breaking, consider the bottom half of an empty wine bottle, a symmetrical upward dome with a gutter for sediment. If a ball is poised exactly at the peak of the dome, the situation is symmetrical. But the ball may spontaneously break this symmetry and roll down the dome to some point in the gutter, a point of the lowest energy. The overall symmetry has spontaneously broken, although the bottle and the ball retain their individual symmetries. Applied to any given physical theory, this idea implies that a *particular solution* to the equations of the theory can be less symmetrical than the theory itself. Such notions underlie the electroweak theory and the theory of strong interactions, two recent theoretical triumphs of modern physics.

Another discovery of our century is that the laws of physics are stunningly general. Thus, notions of symmetry can be applied even to theories about how the universe came to be the way it is today. If one examines modern cosmological theories (for example, those of inflation and the hot big bang) one sees that, as a function of decreasing temperature and increasing time, the universe evolved to give rise to fundamental particles by symmetry-breaking events. At some time far into this process (and long after the particles and fields we know were formed), galaxies, stars, and our solar system with its planets appeared. On Earth, by a process still unknown in detail, life originated and evolution occurred, leading eventually to the emergence of mind. How Fritz Lipmann would have loved to have known what that process was!

What may we offer as a new principle underlying the evolutionary development of mind and intentionality in this set of events? I submit that the new principle is one of memory, one that takes many forms but has general characteristics that are found in all its variations. I am using the word "memory" here in a more inclusive fashion than usual. Memory is a process that emerged only when life and evolution occurred and gave rise to the systems described by the sciences of recognition. As I am using the term memory, it describes aspects of heredity, immune responses, reflex learning, true learning following perceptual categorization, and the various forms of consciousness (figure 20–2).

In these instances, structures evolved that permit significant correlations between current ongoing dynamic patterns and those imposed by past patterns. These structures all differ, and memory takes on its properties as a function of the system in which it appears. What all memory systems have in common is evolution and selection. Memory is an essential property of biologically adaptive systems.

This extension of the term may seem hopelessly broad. But let us see what all these phenomena have in common, for it is actually quite specific. What they have in common is *relative stability of structure under selective mapping events*. To make myself clear, I shall say something here about structure, stability, and mapping. The physical law concerned with structure and stability is the second law of thermodynamics. This law states that entropy, a measure of the disorder of a system, must spontaneously increase or remain the same but never decrease in a closed system. (By a closed system I mean one in which energy and matter neither enter nor leave.) The most orderly possible system is that of a perfect crystal (one with absolutely identical spacing of its atoms in a symmetrical lattice) at a temperature of absolute zero.

Since the earliest evolution of the universe its entropy has been increasing. But in parts of the universe that are *open* systems (ourselves, for example), entropy can actually *decrease* locally as a result of the transfer of matter and energy. Various chemical interactions give rise to stable structures, including those of molecules in living forms. The stability of structures and their energetic transactions are governed by the laws of thermodynamics, including the second law. It is now clear that stable chemical structures can exist in the absence of life or living forms. Indeed, even in outer space, evidence has been found of organic molecules that are similar to those in our bodies—molecules formed by the collision of nitrogen, oxygen, and carbon, for example. The conditions of their formation and dissolution, of their stability, are determined by energy and entropy.

However stable these molecules may be, they lack a hereditary principle. They do not show any ability to replicate themselves—to make molecules that might be called their progeny by using their own structure as a template. I want to be clear about how I am using the word stability in connection with memory. After all, periodic crystals exist in nonliving domains (for example, in rocks) that add atoms to their structures to become larger, following the same rules of symmetry. Such crystals do not *replicate; they grow*. What is the difference?

TYPES OF MEMORY

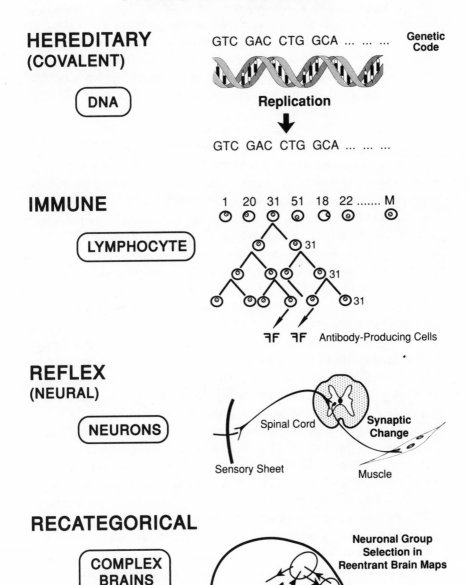

HEREDITARY
(COVALENT)

GTC GAC CTG GCA ... ...   **Genetic Code**

(DNA)

Replication

GTC GAC CTG GCA ... ...

IMMUNE

(LYMPHOCYTE)

1   20   31   51   18   22 .... M

31
31
31

ꟻF ꟻF   Antibody-Producing Cells

REFLEX
(NEURAL)

(NEURONS)

Spinal Cord   **Synaptic Change**

Sensory Sheet

Muscle

RECATEGORICAL

(COMPLEX BRAINS)

Neuronal Group Selection in Reentrant Brain Maps

FIGURE 20–2
Some kinds of memory.

In a replicating system, there is an *aperiodic* structure that undergoes a kind of mapping; think of the sequences of DNA in chapter 6. A chemical reaction faithfully copies the aperiodic structure, resulting in daughter structures. But this fidelity is not absolute; mutant structures arise that are also copied. The result is a variant population. Finally, the stable aperiodic structure *maps* through additional chemistry to make other kinds of structures that contain it, so that favorable variants have a selective advantage when further copies are made.

This abstracted description corresponds to that of a living system: a self-replicating system undergoing natural selection (see chapter 6). The aperiodic structure is DNA or RNA, and the container to which they map is made of various protein products. But notice that the main process, which is lacking in nonliving forms, is the hereditary principle. Notice also that this hereditary principle, which allows an increase in the population of favored variants over time, depends on the stability of chemical bonds. In the case of DNA, these are the covalent bonds linking different nucleotide bases together to make a genetic code of linked triplet codons, each one corresponding to one of twenty amino acids that make up protein chains.

The energy and entropy conditions in the temperature range under which life flourishes assure that a hereditary process takes place. But it is historical *selection* events that result in the actual sequences found in the population.

The appearance of this hereditary process is a new kind of event—a form of memory. Aside from variations introduced into the sequence that have proven favorable, it is the ability to retain much of the order or mapping of the parent aperiodic structure that enables these systems to continue. They have stability of structure under selective mapping events. But notice that this "memory" is not perfect (as it must be, by contrast, in computer messages). Indeed, to some degree, it *must* contain errors (changes in entropy) or mutants for the system to be a selective one—to be one that is able to respond adaptively to unforeseen environmental events because there is population variance.

As these structures evolved and cellular populations formed into animals with many linked cells and with nervous systems, a new kind of memory appeared. This occurred as a result of synaptic changes in the nervous systems of these animals. Because of neuronal group selection, behaviors that proved adaptive were stabilized by selection within a single animal's lifetime. Memory based on synaptic change is essential for such behaviors.

In vertebrates, the requirement that their immune systems make the distinction between self and nonself resulted in the selection of individuals who had a variant of the gene for the neural cell adhesion molecule, N-CAM. By introducing somatic variation into what were to become immunoglobulin (antibody) molecules and by combining that process with the faithful replication of cells selected by foreign molecules, a new recognition system appeared (see figure 8–1). This system had immune memory: The selection of lymphocytes by antigens led to changes that were retained for the entire lifetime of the individual.

Yet again, the evolutionary elaboration of sensory receptors and motor sheets in animals with increasingly sophisticated brain maps made memories based on perceptual categorization possible. With the appearance of conceptual capabilities and even more sophisticated mapping, synaptic change in response to novelty occurring within populations of neuronal groups led to additional kinds of memory.

Each memory reflects a system property within a somatic selection system. And each property serves a different function based on the evolution of the appropriate neuroanatomical structure. These higher-order systems are selective and are based on the responses to environmental novelty of populations of neuronal groups arranged in maps. They are recognition systems.

At some transcendent moment in evolution, a variant with a reentrant circuit linking value-category memory to classification couples emerged. At that moment, memory became the substrate and servant of consciousness. With the emergence of language in the species *Homo sapiens*, the iteration of this same principle in specialized linguistic memories made higher-order consciousness possible. And within culture, higher-order consciousness eventually gave rise to a scientific description of nature, one that allows us to study the origins of our own existence in the universe.

This description of the development of memory is so different from the previous one describing the development of the cosmos following symmetry principles as to seem incommensurate with it. The biological story is a *local* saga so far told only on Earth: It is historical, it occurs in a very narrow temperature range, it is extraordinarily complex and specific to particular structures, it takes unexpected and different forms, and it is dizzying to consider in detail. But the saga begins in a world governed by symmetry. Only with symmetry *breaking*, only with the formation of chemistry, only with the appearance of large, stable molecules, only with the appearance of irreversible selection events, only with the evolution of means described by the sciences of recognition, could memory lead to the appearance of mind. Symmetry principles govern the possibility that memory can arise, but only after symmetry breaking occurred, leading to chemistry and to living and evolving organisms, could memory develop.

Memory underlies meaning. With the transformation of meaning made possible by the embodiment of concepts as described by the TNGS, it became possible within human cultural history to develop true information-processing systems. The historical development of science by social transmission in human culture has made it possible for us to loop back into the truth through chains of knowledge (see figure 14–1). But unlike the development of memory, this explosive transmission is no longer Darwinian. It follows Lamarckian rules because of the character of informational systems and the nature of meaning itself. The contents of informational systems are transferred by *use*; no genetic hereditary principle is needed. The transfer is to somatic systems, each one unique, and the results have been stunning—the transformation of the environment by the human mind in ways that are both valuable and horrendous. They are results that should inspire caution as well as deserving pride.

I have tried in this book to develop a view of the mind based on scientific evidence. Given the state of our knowledge, this view must remain speculative. Although it has philosophical consequences, it is still basically a scientific view, subject to disconfirmation. Despite our ignorance of many of the detailed workings of the brain, I believe it is important to encourage the formulation of such views now. One of them will help point us in the right direction for a while. This is the most one can expect of a scientific theory, beyond the understanding it provides us. A theory exists so that we may build a better theory.

Late as it is in this journey, it will not hurt to stress again that what I have been concerned with here *is* a theory—and not an accepted one. It remains to be tested rigorously, and I have proposed ways of doing so in my previous trilogy on morphology and mind. Like all the theorists I have known, I believe that my theory is correct until proven otherwise. The unit of selection in successful theory creation is usually a dead scientist. One hundred of us go to our graves certain that we are right, but only one turns out to be so. Rarer still is a living scientist accorded such acquiescence. But we each must act as if theories are as important as any other scientific pursuit, risky as they are. Hope and belief are as important in science as they are elsewhere; the difference is that in science they must yield to experiment.

The theory of the mind I have put forth here disclaims the possibility of knowledge beyond doubt. This should not disappoint us, given the history of the success of science in the last three centuries. If the future course of science is determined at all by its present reach, we may expect a remarkable synthesis in the next century. But a "theory of everything" will certainly have to include both a theory of the mind and a fuller theory of the observer. Physics and neuroscience will unite in a more complete comprehension of the relation between the principles of symmetry and memory. They will exist in a tense harmony, one that will make it possible to understand not only the world but also human observers and their place in it.

EPILOGUE

I started this book by telling you I thought its subject was the most important one imaginable. This statement is obviously true in the sense that without a mind there is neither a subject (you or I) nor any subject matter. But I hope our trip through the layers and loops—from molecules to mind and back again even to fundamental particles—has persuaded you of another, less obvious aspect of the importance of neurobiology: that without an understanding of how the mind is based in matter, we will be left with a vast chasm between scientific knowledge and knowledge of ourselves.

This chasm is not unbridgeable. But biology and psychology teach us that the bridge is made of many parts. The solution to the problem of how we know, feel, and are aware is not contained in a philosophical sentence, however profound. It must emerge from an understanding of how biological systems and relationships evolved in the physical world.

When that evolution resulted in language, the imaginable world became infinite. There is great beauty and much hope in the realization of this open-endedness of imagination. But we must continually return from that world to the world of matter if we are to see how as conscious observers we are actually placed within our own descriptions. Analyzing that placement will be one of the major goals of the science of the future.

What form this science will take, it would be foolish to predict. It is sufficient and consoling to know that, whatever form it takes, the conscious life it describes will always remain richer than its description.

MIND WITHOUT BIOLOGY: A CRITICAL POSTSCRIPT

No one likes to spend much time being critical when there is creative work to do. But in order to explain why the kind of biological theory put forth in this book is needed, I have to do a bit of bashing—to criticize several received ideas and established points of view. As I stated in the body of this book, a number of prevailing views about consciousness and the mind are simply untenable, however well established they may be. Why bother with them at all? There are two reasons. First, they are dangerously seductive; sooner or later even the uninitiated reader will run into one or another version of them. And second, a critical analysis of these notions helps to define further the nature of our task, which is to show how the mind is embodied.

There is a third reason: wrong as they may be, these views—that strange physics may hold the key, that the brain is a computer, that we have a kind of built-in language machine in our head—are interesting, whatever their deficiencies. But to convey that interest involves presenting some tedious detail and some rather abstract arguments that would have interrupted my descriptions of the biology of the brain. Therefore I have decided to save my critique of these views for this Postscript.

My goal is to dispel the notion that the mind can be understood in the absence of biology. What I am presenting here are not afterthoughts; they are extensions of points made in the body of the book, intended for the experts but also for the curious who may want to know more.

Readers should not be surprised that the discussion encompasses large numbers of disciplines and jumps from one to the next. The hardest to grasp are perhaps cognitive science and linguistics, both abstract multidisci-

211

plinary areas. But once the obstacles are cleared, they are also fascinating and enormously challenging. Before taking them up, let us turn again to physics.

PHYSICS: THE SURROGATE SPOOK

A spook is a specter or ghost, a disembodied spirit that haunts or scares you. It must seem strange that I am calling that most rationally based of sciences, physics, a surrogate spook. But this is what it becomes when it is applied directly to the mind. Let me explain what I mean.

One way out of the dilemma imposed by the embodiment of mind and the apparent mysteries of consciousness is to make mind and consciousness direct properties of matter. In its most extreme form, this becomes the philosophical doctrine called panpsychism. Panpsychism proposes that all matter, even the finest particles, is a bit conscious, or even that the whole universe is conscious. After all, the reasoning goes, we want to be able to say that mind and matter are connected. If we get a sufficient number of very slightly conscious particles together in the right way, the end result is a conscious human being. This view does not say how one would determine that a particle is conscious, much less a human being.

Such a position "scientizes" another view originally based on the philosophy of idealism. In it the world is perceived only through the mind and thus perhaps, as Bishop Berkeley proposed, there is no matter, only mind. On hearing this, Dr. Johnson kicked a stone and stated, "I refute it thus." A better refutation comes from the theory of evolution: If natural selection gives rise to sentient animals, it is difficult to see how the selecting environment *and* the brain can both be mental events in a single sentient animal that also has progeny undergoing natural selection. The mind reels trying to comprehend how such a complication would ever come to pass.

The theory of natural selection did just as much damage to Plato's idealist notion of essentialism—that there is a world of perfect essences of which the exemplars in the actual world are merely flawed examples. Species are not essences or types; they are the result of selection from variation.

Some very intelligent people have been attracted to panpsychism, idealism, and essentialism. One was the Irish poet William Butler Yeats, who wrote the mystical tract "A Vision" and some extraordinary poems reflecting his thoughts on occult matters. Brains and intellectual gifts are no guarantee against attraction to the spooky and mystical. Under some circumstances it is consoling to have such beliefs, particularly if one clings to notions of immortality. But as my mother said as she lay dying, "I'm in no hurry." When asked why, she smiled and said, "Because no one has come dancing back to tell me what a good time they've been having."

Most good physicists are hardly committed to notions of panpsychism or disembodied spirits. But some very good physicists have nevertheless reached beyond the biological facts and have supposed that the answers to the riddle of consciousness, for example, will reside in a new theory of physics, such as a theory of quantum gravity. To explain why they might be tempted to do so, and why I think they are simply providing us with a surrogate spook, I have to say a few more words about the differences between physics and biology.

Physics is the mother of all the sciences: the earliest, the most general in its reach, the most fundamental. It differs from biology in its generality: it applies equally well to all intentional objects (including human beings) and nonintentional objects. In contrast, biology as we know it is specific. It concerns happenings taking place within a very narrow range of temperature (or energy), pressure, and chemistry. Even more specific is the fact that biology is historical. Evolution is based on a *particular* historical sequence of natural selection from populations of variant organisms. Nothing of the sort has to be considered in formulating the general laws of physics.

This century has seen an astonishing intellectual revolution based on Planck's finding that energy is radiated from matter in finite, discrete packets, or quanta, and on Einstein's theory of relativity, which replaced space and time with the notion of spacetime and advanced the notion of gravity and matter as representing the curvature of a four-dimensional spacetime manifold. The elaboration of these revolutionary ideas led to changes in our ideas of measurement (figure P–1) and radically challenged ordinary notions about the simultaneity of events and about causation. These ordinary notions were replaced by strange, or at least unfamiliar, ones. The elaboration of Planck's and Einstein's work also led to some extraordinary problems that remain unsolved to this day. Their "strangeness" has tempted some scientists to tuck the problem of consciousness in with them.

The ideas behind these basic physical laws can indeed be strange (read "unfamiliar," in the sense of "not commonsensical"). Unlike the ideas of biology, they are *very* general and are often best expressed in mathematical

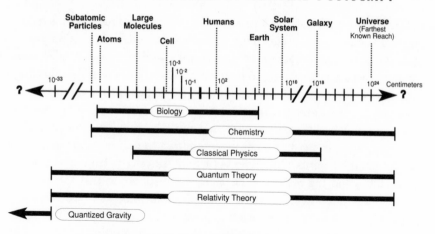

Scales of Nature and Applicable Theories

FIGURE P–1

Scales of nature as established by physics. Below 10^{-33} cm, the theory of relativity and quantum theory do not hold. At the level of molecules and below, quantum theory is essential (and of course it applies at all scales above 10^{-33} cm). At very high velocities and accelerations one must apply relativity theory. But at the level of macroscopic objects (including humans and their brains), one may approximate nicely with classical descriptions; the convergence of quantum and classical theories at large scales is known as the correspondence principle. Note that the size range of brains and the temperature range of living things are both quite narrow. The scale is in powers of ten—that is, it is logarithmic.

theories that have great power and beauty. One example is that of symmetry, which I discussed in chapter 20.

Such notions of physics, with their generality and predictive power, are beguiling. As powerful as they are, however, profound problems arise in understanding their application. An example comes from the theory of quantum measurement, which must be taken into account when one attempts to measure the position or momentum of a fundamental particle. In facing the paradoxes that arise from these attempts, distinguished mathematicians such as John von Neumann and equally distinguished physicists such as Eugene Wigner were tempted to propose that consciousness itself causally intervenes in the process of quantum measurement.

There are many issues related to these proposals, and to discuss them all here would take us too far afield. But let me sketch out one aspect of the quantum measurement problem to show *why* these scientists were tempted to bring consciousness into physics. As I do so, however, please keep in mind that physics is concerned with formal correlations of the most general properties of things in spacetime. Theories of physics are *not* concerned with the senses proper, with categorizing nameable macroscopic objects, or with intentionality. If one delves into quantum mechanics, it is easy to forget these restrictions, because the decisions of the observer appear to affect the measurements he or she makes. To understand this, we have to consider a few salient features of quantum theory.

Quantum theory is the most generally applicable of all theories. In dealing with enormous energies and very small particles, this theory has revealed behavior that confounds ordinary expectations. For example, one particle cannot be identified as distinguishable from another. Particles show duality of behavior: Under one set of circumstances they are best described as waves, in others as particles. Indeed, as Max Born first suggested, the fundamental wave function ψ in the Schrödinger wave equation, when taken as an absolute value and squared, is a measure of the probability of finding a particle in a given position of space—anywhere!

If one attempts to determine that position in an experimental setting, however, one loses forever the possibility of determining the momentum to the same precision. This so-called Heisenberg uncertainty is fundamental; there is a conjugate relation between the position and the momentum (the mass times the velocity) of a particle, and this relation sets the precision of the product of these variables to a value no less than Planck's constant. This is *not* just because to measure a particle's position precisely one must use particles or waves of much smaller wavelength and thus of higher energy, inevitably "kicking up" the particle's momentum. It is a *fundamental* property of the theory. In considering this relationship operationally, one begins to get a feeling for the strange flavor of quantum theory. If one (the physicist observer) chooses to measure the position of a particle to a certain precision, the act of setting up and carrying out the measurement precludes forever and irreversibly the measurement of the momentum to a similar precision. According to the theory, however, no bias exists before the measurement: The wave function ψ is a linear combination of functions describing all possible outcomes of the measurement, and when a measurement is made the wave function "collapses" or "projects onto" one of the possible outcomes.

As von Neumann pointed out, the macroscopic measuring instrument is also described by a quantum mechanical wave function (practically speaking, we do not need quantum theory to describe such objects physically). He then formally showed that one cannot draw a line from the wave function of the particle all the way up to the act of the observer to establish the value of ψ at any scale. The "collapse of the wave function" is determined just when the apparatus and the particle interact to give a definite measurement. This collapse was attributed by Wigner to be the result of the intervention of the observer's consciousness. After all, the observer decides to set up the apparatus, decides whether he or she is interested in position or momentum, and actually makes the measurement! To determine the state of this apparatus in von Neumann's view, one apparatus needs another, and that one needs another one, and so on, regressing in an infinite fashion. In Wigner's scheme a phenomenon only becomes actual (that is, the regress is ended) when the observer becomes conscious of it.

In all fairness it should be said that other distinguished physicists have interpreted the quantum measurement problem without calling the consciousness of the observer into play. Niels Bohr, the father of quantum theory, declared that there is no ultimate or deep reality; one simply applies the principle of complementarity (of which Heisenberg's principle is perhaps the most elegant expression) and then obtains the result dictated by the *entire* situation of measurement, particle, apparatus, and observer. Bohr's "Copenhagen interpretation" is the position taken by most physicists who use the theory. It gives a formula describing what one observes with an apparatus, one that is ultimately made up of the same kind of quantum particles one is measuring. Other physicists have even proposed that there is no "collapse" of the wave function. Instead they conceive that there are "many worlds," in each one of which the function takes on a possible value alternative to the one in this world with this observer whom we see here and now. Still others have proposed a "quantum potential" that might even involve faster-than-light signaling, something that contradicts Einstein's theory of relativity!

I once discussed this problem at lunch with my friend Isidor Rabi, a great physicist, just five months or so before his death. He looked at me with a puckish smile and said, "Quantum mechanics is just an algorithm. Use it. It works, don't worry." I nagged him, saying "Rab, don't tell me you are getting like Einstein, dubious of the whole thing." He replied with a laugh, "Listen, if I'm having trouble with God, why shouldn't I have trouble with quantum mechanics?"

This brings us to the issue at hand: with such strangeness, why not get a little stranger and propose that additional, as-yet-undiscovered physical fields or dimensions might reveal the true nature of consciousness? This is a subtle but also more off-putting way of proposing physics as a surrogate spook. A good example is the position taken by the mathematician and cosmologist Roger Penrose in his wide-ranging book *The Emperor's New Mind*, which takes as its theme the nature of consciousness. The book abounds in clear examples of paradoxes in physics and in descriptions of the axiomatic limitations of mathematics. On intuitive grounds alone, based on his personal experience as a mathematician and in appreciation of these axiomatic limits, Penrose rejects the notion of the brain as a computer. He points out the limits of quantum mechanics and relativity in domains where the dimensions are so small (below the so-called Planck length of 10^{-33} cm) that such theories cannot apply. And he calls for a theory of quantum gravity that would extend these theories. Then, by a remarkable leap, he proposes that the mystery of consciousness will be resolved when a theory of quantum gravity is satisfactorily constructed.

I suppose the reasoning is as follows: The decisions of the observer and the operator *are* intimately involved in quantum mechanical and relativistic measurements. The observer's mind constructs and applies mathematical theories, the statements of which transcend the ability of formal axiomatics to prove or disprove them, yet he or she can check their truth and meaning as a computer cannot. Like everything else, the observer's brain is ultimately made up of particles obeying quantum laws, particularly at its synapses, where most of the action is. Physical laws as currently formulated do not account for consciousness. Neither can they explain quantum gravity. Perhaps an explanation of quantum gravity will provide the clue to consciousness, which seems to hover around all of our theories!

Truly, this is physics as the surrogate spook—more reasoned, perhaps, than many a spook in religious tracts or in occult accounts, but no more rewarding in the end. Indeed, while Penrose's book contains many fine descriptions of physics, it bears little on the problem of consciousness as intentionality, for it ignores both the psychological and the biological knowledge essential to understanding the problem. Penrose's account is a bit like that of a schoolboy who, not knowing the formula of sulfuric acid asked for on an exam, gives instead a beautiful account of his dog Spot.

What is missing from his and other accounts is a sober scientific analysis of the proximate structures and functions related to awareness: an account of real psychology, of real brains, and of their underlying biology. While physics obviously provides the necessary bases for biology, it doesn't concern itself with biological structures and processes and principles. These are quite special and much more demonstrably connected with the mind than are the general ideas of symmetry and quantum measurement, important as these are to understanding the existence of all things. Indeed, it is a much more sensible thing to construct and test a theory of mind based on biological processes than to postulate exotic physics as an explanation. There is plenty of direct evidence, after all, for anatomy affecting consciousness.

Until we reach a biological impasse, therefore, we would do well to reject as a category error the notion that exotic physics itself will give a description of the observer's consciousness. We must not confuse the bases of the workings of our minds with our minds' fine intellectual constructions, such as theories of physics. (An irreverent description of a horse show may focus our attention on the nature of the category error: a horse show is a bunch of horses showing their asses to a bunch of horses' asses who are showing their horses.)

We must nonetheless be grateful to Penrose, a great scientist, at the very least for drawing renewed attention to an even more commonly made category error: that which assumes the brain is like a computer. Let us turn to it, for its consideration brings us much closer to the fundamental issue than does any further consideration of physics itself.

DIGITAL COMPUTERS: THE FALSE ANALOGUE

If physics won't do as a surrogate spook, what about an unusual physical object or construct—the digital computer? After all, this most remarkable of all inventions of the twentieth century seems to carry out a remarkable number of functions that at first glance appear to be mindlike.

Extraordinarily silly things have been proposed about the capacity of machines to think. For the most part, the silliness arises from the analogy between thinking and logic. The indisputable fact is that computers carry out logical operations. The rub is that logic alone carried out on a computer no more constitutes thinking than the physical events of adding up numbers on an abacus resemble what goes on in the brain during the performance or creation of arithmetic by a mathematician.

To see why this confusion has arisen, I must explore a bit of the theory behind the digital computer. That theory owes itself largely to the work of the late Alan Turing, a British mathematician who committed suicide by biting a poisoned apple. As a discovered homosexual, he had been given a forced choice by the British courts either to go to jail or to take the feminizing hormone estrogen. He chose the latter, with feminizing effects on his body, and who knows what effect on his brain. That brain gave rise to a powerful set of mathematical ideas, one of which is known as a Turing machine.

Turing defined an abstract class of automata and showed that any member of that class can compute any of a large class of functions. (All but a few special-purpose computers are Turing machines.) A Turing machine (figure P–2) is a finite-state machine with an infinite tape; in a given square on the tape it can write either a 0 or a 1, and it can shift the tape one square (containing one such digit) to the left or to the right. It has instructions

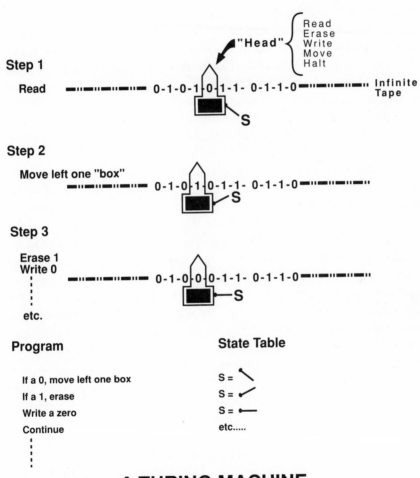

A TURING MACHINE

FIGURE P–2

A Turing machine. This abstraction has been shown to represent the functional operations of practically all computers. Turing's analysis holds for real-world computers even though a Turing machine (unlike a real-world computer) has to go through many more steps than is convenient to carry out a simple information processing procedure or algorithm. The idea is a triumph of clear reasoning.

containing conditions and actions, and it carries out an action if a particular condition is satisfied. The condition is determined by the symbol on its tape under the tapehead and by the state of the machine, and a given action is any one of the four described above, after which it shifts to the next state specified by the program. A "universal Turing machine" can simulate any particular Turing machine. (Particular Turing machines can have different mechanisms and parts, as long as they obey Turing's description.)

Now comes the temptation to commit a category error. A persuasive set of arguments states that if I can describe an effective mathematical procedure (technically called an algorithm; see figure P–4), then that procedure can be carried out by a Turing machine. More generally, we know that *any* algorithm or effective procedure may be executed by any universal Turing machine. The existence of universal machines implies that the *mechanism* of operation of any one of them is unimportant. This can be shown to be true in the real world by running a given program on two digital computers of radically different construction or hardware design and successfully obtaining identical results (see figure P–3).

On the basis of these properties, the workings of the brain have been considered to be the result of a "functional" process, one held to be describable in a fashion similar to that used for algorithms. This point of view is called functionalism (and in one of its more trenchant forms, Turing machine functionalism). Functionalism assumes that psychology can be adequately described in terms of the "functional organization of the brain"—much in the way that software determines the performance of computer hardware. Functionalism is concerned not only with functions performed by various systems but also with the relations between their components, particularly as they cause other relations to take place. Functionalist theories are indifferent to the mechanical instantiation of a system, and thus they deal in abstract terms with such relations.

In the functionalist view, what is ultimately important for understanding psychology are the algorithms, not the hardware on which they are executed. According to functionalism, what the brain does may be adequately described by algorithms. Furthermore, the tissue organization and composition of the brain shouldn't concern us as long as the algorithm "runs" or comes to a successful halt (figure P–4). (This "liberal" position affirming the absence of any need for particular kinds of brain tissue suffuses much of present-day cognitive psychology.)

If we accept this position, an analysis from formal logic known as Church's thesis suggests that if any consistent terminating computational method exists to solve a given problem, then a method exists that can run on a Turing machine and give exactly the same results. For problems that

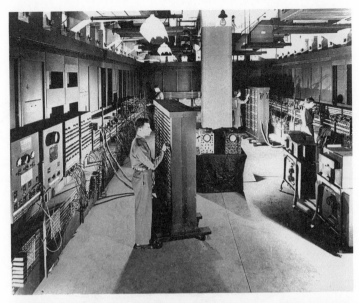

FIGURE P–3

Two computers in the real world. Top: ENIAC, the first practical digital computer.
Bottom: An N-CUBE, a commercially available supercomputer based on massively
parallel processing. ENIAC filled a large room and carried out about five thousand
instructions per second; the N-CUBE is about the size of a standard office desk and carries
out nearly eight billion instructions per second. If you have the money, I'll reduce the time,
but the principle (Turing's) is in any case the same.

221

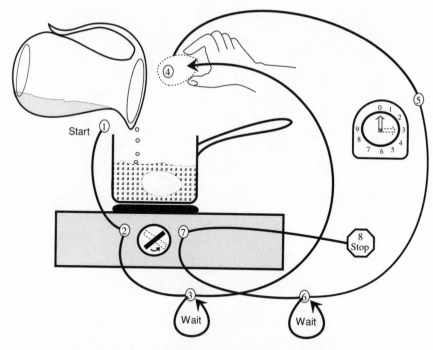

AN ALGORITHM FOR BOILING AN EGG

① Put water in pot

② Turn on heat

③ If water is not boiling go to step ③
otherwise continue to step ④

④ Put egg in water

⑤ Set timer to 3 minutes

⑥ If timer has not gone off go to step ⑥
otherwise continue to step ⑦

⑦ Turn off heat and cool

⑧ Finished ... retrieve, peel, and eat egg

FIGURE P–4

An algorithm for boiling an egg. One for adding two numbers would have equally explicit instructions.

can be solved consistently in a specified finite amount of time, a Turing machine is as powerful as any other entity for solving the problem, *including the brain.* According to this analysis, either the brain is a computer, or the computer is an adequate model or analogue for the interesting things that the brain does.

This kind of analysis underlies what has become known as the physical symbol system hypothesis, which provides the basis for most research in artificial intelligence. This hypothesis holds that cognitive functions are carried out by the manipulation of symbols according to rules. In physical symbol systems, symbols are instantiated in a program as states of physical objects. Strings of symbols are used to represent sensory inputs, categories, behaviors, memories, logical propositions, and indeed all the information that the system deals with. The operations needed to transform strings of input symbols into strings of output symbols are computations, and according to the physical symbol system hypothesis, they may therefore be carried out by any suitably programmed Turing machine. As I have said, these operations are purely formal in nature; that is, they may be carried out without reference to the *meanings* of the symbols involved. (As we saw in chapters 2 and 12, a set of these rules is known as a syntax.) The particular design of the computing device responding to these syntactical rules is of concern only in that it must meet certain requirements of speed and memory capacity in order to be able to complete its work in a reasonable amount of time.

Why won't this position do? The reasons are many, but before I take them up, remember the connection of physical symbol systems with the argument for functionalism (which has many variants, all of which share the formal causal position). If any of the forms of functionalism is a correct theory of the mind, then the brain is truly analogous to a Turing machine. And in that case, the relevant level of description for both is the level of symbolic representations and of algorithms, not of biology.

Not all forms of functionalist theory impose such a large degree of identification of processes in the mind with processes in Turing machines. The strongest position, originally formulated by Hilary Putnam and known as "Turing machine functionalism," postulates that the two are entirely equivalent. This view is no longer widely accepted; indeed, it has been rejected by Putnam himself. Weaker forms of functionalism do not require a strict equivalence between brain states and Turing machine states. However, *all* forms of functionalism hold that two systems having isomorphous functional states must be in identical cognitive states, irrespective of any differences in their physical makeup. This conclusion is a close cousin to some of Turing's results on universal computation. These results amount to the assertion that two computers having identical abstract state transition tables and identical symbols on their tapes (see figure P–2 for definitions and examples) are carrying out the same computation, regardless of the physical form taken by the processor and the tape.

And now the coup de grace (actually multiple coups)! An analysis of the evolution, development, and structure of brains makes it highly unlikely that they are Turing machines. As we saw in chapter 3, brains possess enormous individual structural variation at a variety of organizational levels. An examination of the means by which brains develop indicates that each brain is highly variable. Indeed, a simple calculation shows that the genome of a human being (the entire collection of an individual's genes) is insufficient to specify explicitly the synaptic structure of the developing brain. Moreover, each organism's behavior is biologically individual and enormously diverse, whether or not that organism registers or reports subjective experiences as human beings can.

More damaging is the fact that an analysis of ecological and environmental variation and of the categorization procedures of animals and humans (which I discuss in the next section) makes it unlikely that the world (physical and social) could function as a tape for a Turing machine. Arguing along similar lines, Putnam has repudiated his original and other dependent models of functionalism. His central point is that psychological states including propositional attitudes ("believing that p," "desiring that p," and so on) cannot be described by the computational model. We cannot individuate concepts and beliefs without reference to the environment. The brain and the nervous system cannot be considered in isolation from states of the world and social interactions. But such states, both environmental and social, are indeterminate and open-ended. They cannot be simply identified by any software description. Functionalism, construed in this context as the idea that propositional attitudes are equivalent to computational states of the brain, is not tenable.

Another philosopher, John Searle, has also been a strong critic of the functionalist position. His opposition is based on the idea that no purely computational specification provides sufficient conditions for thought or for intentional states. His argument (which applies to higher-order consciousness, the kind we have as humans) is that computer programs are defined strictly by their formal syntactical structure, that syntax is insufficient for semantics, and that in contrast, human minds are *characterized* by having semantic contents. Semantic contents involve meanings, and a syntax does not in itself deal with meanings. The rejection of functionalism in this position is obvious. Moreover, Searle maintains that, inasmuch as consciousness is identified in humans with a type of intentionality that is inexorably accompanied by subjective experience, by definition no organism can have intentional states if it lacks subjective experience. Computers lack such experience. Certain functionalists (possibly the majority) restrict their claims to statements that preclude subjective or phenomenal properties. Given his arguments, Searle would reject their claims (rightly, I think) as having no bearing on the origins of consciousness or thinking.

What is at stake here is the notion of meaning. Meaning, as Putnam puts it, "is interactional. The environment itself plays a role in determining what a speaker's words, or a community's words, refer to." Because such an environment is open-ended, it admits of no *a priori* inclusive description in terms of effective procedures. Moreover, we have seen in this book that the actual body of the speaker plays an equally great role in determining meaning. Arguments concerning semantics and meaning are important for any theory of consciousness (and thinking) that takes as its canonical reference our own phenomenal experience as humans and our ability to report that experience by language.

Notice the difference when we turn to computers. For ordinary computers, we have little difficulty accepting the functionalist position because the only meaning of the symbols on the tape and the states in the processor is *the meaning assigned to them by a human programmer.* There is no ambiguity in the interpretation of physical states as symbols because the symbols are represented digitally according to rules in a syntax. The system is *designed* to jump quickly between defined states and to avoid transition regions between them; electrically, each component always goes to a "zero" or a "one." The small deviations in physical parameters that do occur (noise levels, for example) are ignored by agreement and design. One purpose of all these conventions is to assure that any differences between two systems that occur because of their different ways of physically representing symbols should indeed have no meaning. Different hardware is not an issue as long as the hardware performs. Remember, though, that this portability of functionalist systems across different hardware implementations is bought at the price of requiring primitive functional processes to operate on symbolic representations of information.

Now we begin to see why digital computers are a false analogue to the brain. The facile analogy with digital computers breaks down for several reasons. The tape read by a Turing machine is marked unambiguously with symbols chosen from a finite set; in contrast, the sensory signals available to nervous systems are truly analogue in nature and therefore are neither unambiguous nor finite in number. Turing machines have by definition a finite number of internal states, while there are no apparent limits on the number of states the human nervous system can assume (for example, by analog modulation of large numbers of synaptic strengths in neuronal connections). The transitions of Turing machines between states are entirely deterministic, while those of humans give ample appearance of indeterminacy. Human experience is not based on so simple an abstraction as a Turing machine; to get our "meanings" we have to grow and communicate in a society.

The abstract beauty of Turing machines is beguiling. But one must watch out for excessive abstraction even in science, where it usually lends great power to much of our thinking. Abstraction in some contexts is quite foolish. The story is told of a race track that was failing financially. The track management consulted three experts: an accountant, an engineer, and a physicist. The accountant recommended restructuring of the balance sheets; the engineer suggested that the slow track could be fixed by slight banking and better drainage. When his turn came, the physicist went up to the blackboard, drew a circle, and said, "Let us replace the horse with a sphere."

In contrast to computers, the patterns of nervous system response depend on the individual history of each system, because it is only *through interactions with the world* that appropriate response patterns are selected. This variation because of differences in experience occurs between different nervous systems and within a single system across time. The existence of extensive individual variation in cognitive systems (see chapter 3) negates the fundamental postulate of functionalism that representations have meaning independent of their physical instantiation. Thus, it would appear that the independence of physical instantiation that is such a prized feature of functionalist systems must be abandoned if a nontrivial level of cognitive performance is to be achieved. (This does not mean that in abandoning the liberal position of functionalism, we must adopt the extreme chauvinist position: that carbon chemistry, wet tissues, and so on, are *absolutely* necessary for cognition to occur. Were that so, the artifacts discussed in chapter 19 could not be constructed.)

Whatever type of internal representations a functionalist system may employ, a procedure is needed for establishing the meanings of the individual units (symbols or their generalizations) and of combinations of units in those representations. It is not easy to see how, in the absence of a programmer, a mechanism could be constructed that would assign meaning to syntactic representations and still preserve the arbitrary quality of those representations, a quality that is an essential part of the functionalist position. But that is our poignant position: we have no programmer, no homunculus in the head.

I could not close here without mentioning that, in recent years, a large amount of work has been done on "connectionist" or "neural network" models of perceptual or cognitive processes. These are formal models in which the connections between network elements are modified in a fashion loosely analogous to synapses. I suppose this justifies the metaphoric "neural," but in other respects the metaphor is strained, as I point out in what follows.

These constructions have been useful in a number of applications. Many of the models begin with assumptions about the nature of intelligent systems similar to those made by workers in artificial intelligence. Unlike classical work in artificial intelligence, however, these models use distributed processes in networks, and changes in connections occur in part without strict programming. Nonetheless, connectionist systems need a programmer or operator to specify their inputs and their outputs, and they use algorithms to achieve such specification. While the systems allow for alterations as a result of "experience," the mechanism of this "learning" is instructional, not selectional. Unlike selectional systems carrying out categorizations on value, the *responses* (not the values) of connectionist systems are specified in advance and are imposed on the system by a human operator under appropriate conditions and with appropriate error feedback to establish the training.

The architectures of neural networks are removed from biological reality, and the networks "function" in a manner quite unlike the nervous system. "Neural nets" use symmetrical and dense matrixlike connections. In general, they do not at all resemble the neuronal structures and the anatomy of which I have written in this book. If neural networks were adopted as the standard model of brain structure and function, we would have to say that they support the view of the brain as a Turing machine. Whatever their interest and usefulness, neural networks are not adequate models or analogues of brain structure. (For readers interested in pursuing these issues, I have placed a reference to two collections of papers on this topic in the Selected Readings at the back of the book.)

Whether digital computers or connectionist models are used as a base, we are left with the same embarrassment. In considering the brain as a Turing machine, we must confront the unsettling observations that, for a brain, the proposed table of states and state transitions (see figure P–3) is unknown, the symbols on the input tape are ambiguous and have no preassigned meanings, and the transition rules, whatever they may be, are not consistently applied. Moreover, inputs and outputs are not specified by a teacher or a programmer in real-world animals. It would appear that little or nothing of value can be gained from the application of this failed analogy between the computer and the brain.

But the field is not abandoned so easily. There remains a large body of work in cognitive psychology based on similar confusions concerning what can be assumed about how the brain works without bothering to study how it is physically put together. Let us turn now to some of the difficulties created in cognitive psychology by one of its central notions: the idea of mental representations.

SOME VICIOUS CIRCLES IN THE
COGNITIVE LANDSCAPE

The blend of psychology, computer science, linguistics, and philosophy known as cognitive science has grown enormously. As with all vigorous efforts, ill-founded or not, much has emerged that is of great interest to scientists and nonscientists alike. Not the least of the positive results has been the routing of simpleminded behaviorism. But at the same time, an extraordinary misconception of the nature of thought, reasoning, meaning, and of their relationship to perception has developed that threatens to undermine the whole enterprise.

To trace the nature of this misconception takes some doing, for it has complex historical, intellectual, and practical roots. I must warn the reader that this field delves into some complicated matters, and I cannot simplify their description altogether. Before I tackle the details, let me give you a short characterization of the misconception. It stems from the notion that objects in the world come in fixed categories, that things have essential descriptions, that concepts and language rely on rules that acquire meaning by formal assignment to fixed world categories, and that the mind operates through what are called mental representations. These representations are supposed by some to be expressed through a language of thought, or "mentalese," as the philosopher Jerry Fodor calls it. Meaning consists of the assignment of symbols in such a language to correspond *exactly* with entities or categories in the world defined by singly necessary and jointly sufficient criteria (classical categories). Thus, a specification of the rules by which representations are manipulated (constituting a syntax), if complete, can be carried out by a computational device. The brain in this view is a kind of computer. (Note the similarity of some of these assertions to those in the last section.)

The acceptance of this view or versions of it is widespread in psychology, linguistics, computer science, and artificial intelligence. It is one of the most remarkable misunderstandings in the history of science. Indeed, not only is it not in accord with the known facts of human biology and brain science, but it constitutes a major category error as well.

We have fooled ourselves in part as a result of our success in removing the mind from nature in the "hard" sciences. The error has been to attribute the characteristics of human mental constructions (such as logic and mathematics) to human reasoning and to the macroscopic world in which we live. Whenever I think of this carving of vicious circles of rational design onto

228

the surface of the cognitive landscape, I cannot help but think of the conversation between two mice in a psychology laboratory. After a successful run in the maze, one mouse says to the other, "You know, I think I have my psychologist trained. Every time I run the maze successfully, he gives me a piece of cheese."

To show you why the ideas of "mentalese," of rules and representations, and of computation will not work, I must take up some of the stricter assumptions of the functionalism underlying cognitive psychology. Then I must consider a view of the world (and particularly the scientific world) called objectivism. Finally, I must look at a central issue: the evidence concerning how we actually categorize the world, both perceptually and conceptually. That done, we will be able to see the errors in reasoning that threaten to undermine the cognitive enterprise. The arguments and data considered here will not be exhaustive; the reader is encouraged to consult the appropriate works in the Selected Readings for more information. I will attempt to sketch the issues in a minimal but incisive manner. They are at the heart of any effort to understand the matter of the mind.

It appears that the majority of those working in cognitive psychology hold to the views I attack here. But there is a minority who hold contrary views, in many ways similar to mine. These thinkers come from many fields: cognitive psychology, linguistics, philosophy, and neuroscience. They include John Searle, Hilary Putnam, Ruth Garret Millikan, George Lakoff, Ronald Langacker, Alan Gauld, Benny Shanon, Claes von Hofsten, Jerome Bruner, and no doubt others as well. I like to think of them as belonging to a Realists Club, a dispersed group whose thoughts largely converge and whose hope it is that someday the more vocal practitioners of cognitive psychology and the frequently smug empiricists of neuroscience will understand that they have unknowingly subjected themselves to an intellectual swindle. The views of this minority will be reflected in what I have to say, but obviously they vary from person to person. The reader is urged to consult these scholars' works directly for a closer look at the diversity of their thoughts and interpretations.

Functionalist Views and the Semantic Representation of Meaning

The central idea underlying much of modern cognitive psychology is that of mental representations. These representations are abstract and symbolic (that is, they stand for some thing or some relation), are formed in a well-defined manner, and follow rules that constitute a syntax. They are supposed to be semantically related to the world by fixed and determinate relationships and by the semantic assignment of symbols to objects in classical categories. They are essential to the formation of "inner models of the world."

The idea of inner models follows the early suggestions of K. J. W. Craik, and in this view, internal representations parallel external structures in the world. The representations are propositional, involving concepts and their relationships, or they are mental images. The origin of images is perception, which in this view is a form of computation, according to the seminal and influential ideas of the late David Marr. The computations that occur on mental structures are governed by a system of rules (or a syntax) and by the representations themselves. The whole system of representation forms a *lingua mentis* or mentalese, a language of thought.

How does this point of view deal with the problem of intentionality? Presumably by declaring that meaning arises from the mapping of *rule-governed* syntactical structures onto *defined* and *fixed* world objects or relations. Such a semantics is exhaustive and determinate and, together with its underlying syntax, it provides a framework for modeling the mind.

How did this functionalist, computational view of the mind, which is highly formal and disembodied, arise? How could anyone accept so abstract a notion of human knowledge, reason, and mental activity? Before I criticize this view of the mind, let us consider the corresponding view of the world that forms one of its foundations.

Objectivism

The term "objectivism" has been used to characterize a view of the world that appears at first sight to be both scientifically and commonsensically unexceptionable. (One analysis I will follow is that of Lakoff; see the Selected Readings.) Objectivism goes beyond the hypothesis of scientific realism, which itself assumes: (1) a real world (including humans but not depending on them); (2) a linkage between concepts and that world; and (3) a stable knowledge that is gained through that link. Objectivism assumes, in addition to scientific realism, that the world has a definite structure made of entities, properties, and their interrelationships (figure P–5). These are capable of definition according to classical criteria of categoriza-

OBJECTIVISM

Objects
(and Events)

Descriptive Terms
(Words and Scientific Theories)

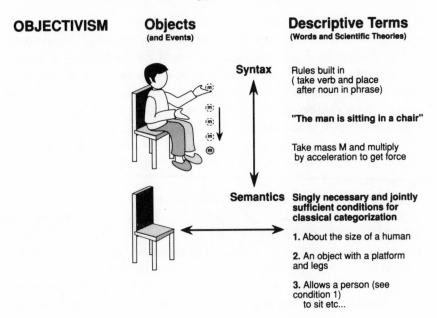

Syntax

Rules built in
(take verb and place
  after noun in phrase)

"The man is sitting in a chair"

Take mass M and multiply
by acceleration to get force

Semantics

**Singly necessary and jointly
sufficient conditions for
classical categorization**

1. About the size of a human

2. An object with a platform
and legs

3. Allows a person (see
condition 1)
    to sit etc...

**MACHINE
FUNCTIONALISM**

Software   **Hardware**

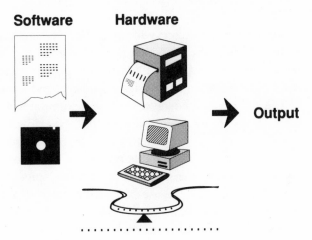

Output

"Information"   **Brain**

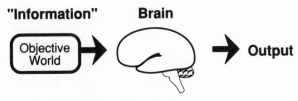

Objective
World

Output

FIGURE P–5
Some aspects of objectivism and functionalism.

231

tion that are singly necessary and jointly sufficient to define each category. The world is arranged in such a fashion that it can be completely modeled by what mathematicians and logicians would call set-theoretical models. These kinds of models, which are seen in mathematical logic, consist of symbolic entities appearing singly or in sets, together with their relationships. Symbols in these models are made meaningful (or are given semantic significance) in a unique fashion by assuming that they correspond to entities and categories in the world. Some of the categorical properties of things in the world are considered to be essential; others are seen as accidental.

Because of the singular and well-defined correspondence between set-theoretical symbols and things as defined by classical categorization, one can, in this view, assume that logical relations between things in the world exist *objectively*. Thus, this system of symbols is supposed to represent reality, and mental representations must either be true or false insofar as they mirror reality correctly or incorrectly. According to objectivism, this correspondence to things in the world gives meaning to linguistic expressions; meaning is based on this "correct" or "incorrect" definition of truth and thought itself is a manipulation of symbols.

This view can certainly be held outside of science. Indeed, the objectivist position seems in accord with much that is commonsensical. But when it is held inside science, it comes close to the Galilean position we discussed in chapter 2. In that sense, human concepts, assertions, and languages are valid only if limited to physics, chemistry, and parts of biology.

We will see that, however sensible it seems at first, this view is woefully incoherent and not in accord with the facts. How then did it arise? Well, one can go a long way with this view in the hard sciences. Removal of the mind from nature *is* a sensible precaution for much of classical chemistry and physics. And many of the major developments in physics have depended strongly on the use of the rigorous formal reasoning at the core of mathematics and logic.

In the late nineteenth and early twentieth centuries, deep investigations into mathematical logic by Gottlob Frege, Giuseppe Peano, and Alfred North Whitehead and Bertrand Russell, followed by the work of Stephen Kleene, Emil Post, Alonzo Church, Alan Turing, and Kurt Gödel, were triumphs of human analysis of the "mechanics" of reasoning by logic. When I was in college, I was enraptured by the elegance of all of this. I spent long evenings with those great dark blue compendia of logical hieroglyphics, the *Principia Mathematica* by Whitehead and Russell. Their very dryness convinced me that I was on the inside track. It was too bad that I had no one to tell me about the human side of these authors at the time. I have since heard that during the writing of these tomes the usually mild-mannered Whitehead once said to his feistier colleague: "Bertie, the world is divided into the simpleminded and the muddleheaded, and I shall leave it to you to decide which one you are." The mathematician G. C. Rota has recently mounted a scathing attack on the excessive reliance on formalism and axiomatics of some philosophers who ape the clarity of mathematics by adopting a symbolic mode of discussion (see the Selected Readings, chapter 14, for a reference to his work).

The subsequent development of the computer, which was partly based on these investigations, reinforced the ideas of efficiency and rigor and the deductive flavor that had already characterized much of physical science. The "neat" deductive formal background of computers, the link with mathematical physics, and the success of the hard sciences looked endlessly extensible. There was a natural tendency to stop a philosophical analysis of scientific exploration at the surface of the human body (the skin and its receptors). Behavior could be analyzed, but not phenomenal experience. In this way, science could remain "extensional," as W. V. Quine put it, and one could declare with him that "to be is to be a value of a variable."

The computational or representationalist view is a God's-eye view of nature. It is imposing and it *appears* to permit a lovely-looking map between the mind and nature. Such a map is only lovely, however, as long as one looks away from the issue of how the mind actually reveals itself in human beings with bodies. When applied to the mind *in situ*, this view becomes untenable.

The difficulties of the computational or mental representational view of mind are manifold and can be usefully grouped into eight types (table P–1). The grouping is not just a matter of convenience but also provides a battle plan for an assault against this view of mind. I recommend that the interested reader consult the Selected Readings to obtain a deeper understanding of the works of the authors mentioned in table P–1. On the assumption that the reader will do so, I treat the issues listed in the table only briefly here. My goal is to sketch the major critical arguments against functionalism and objectivism, not to exhaust them.

Categories: A Crisis for Functionalist Views of Cognition

One of the largest challenges to the functionalist view of mental representations comes from philosophical and psychological work on how we categorize things. Most of this work is concerned with conceptual categori-

TABLE P–1
Some Problems with the Idea of Mental Representation*

---

1. Perception and reason are not governed by classical categories. Biology (particularly Darwin's work) shows essentialism to be false (Rosch, Wittgenstein, Lakoff, Mayr). Similarity is not the same as categorization.

2. Thought is not transcendent but depends on the body and brain. It is *embodied.* Meaning arises from relations to bodily needs and functions. The mind does not mirror nature (Putnam, Millikan, Langacker, Lakoff, Johnson, Searle, Edelman).

3. Memory cannot be described by internal codes or syntactic systems. Moreover, one needs a self and higher-order consciousness to account for its full linguistic manifestation (Searle, Shanon, Gauld, Edelman).

4. Language is acquired by interacting with others in learning events, which initiate the formation of connections between semantics and phonology. It depends on having conceptual systems and values already in place (Pinker, Johnson, Edelman).

5. Minds create their own versions of reality by social and linguistic interactions, and reality, like biology itself, depends on historical events (Searle, Putnam).

6. Computation is not only disembodied; it cannot by itself provide a meaningful relation between symbols and world entities (Searle).

7. Cognition gets its content from the identification of proper functions in a system that depends on evolutionary history. Each part of a proper function has a "normal" explanation, which tells how that system has managed historically to perform that function. "Meaning rationalism," the assignment of meaning from above, is untenable (Millikan).

8. The structure, function, and diversity of the nervous system, as well as its evolution and development, are incompatible with the functionalist view (Edelman).

**The names in parentheses are those of authors whose works appear in the list of Selected Readings included at the end of the book. Please consult these readings for the extended arguments.*

zation in humans, although some of it also concerns perceptual categorization in both humans and animals. The single most striking conclusion arising from a variety of analyses and studies is that people do not categorize things or events in terms of classical categories. Classical categories are those in which membership is defined in terms of singly necessary and jointly sufficient conditions (figure P–5).

Wittgenstein had some of the earliest critical thoughts on this subject. He reflected on family resemblance, noting that category members can be related to each other even if some members do not have any of the properties that classically define the category in common (figure P–6, right). (Imagine that there are *n* properties distributed among the members

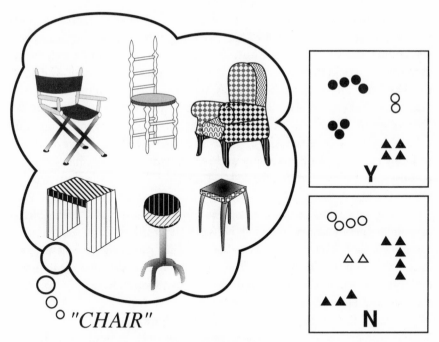

FIGURE P–6

Categorization and polymorphous sets. **Left:** *Chairs are not* necessarily *characterized by singly necessary and jointly sufficient criteria (classical categories).* **Right:** *A polymorphous rule for set membership when classical categorization does not apply. Members of the exemplary set (group marked Y for "yes") have any two of the properties roundness, solid color, or bilateral symmetry. Nonmembers (group marked N for "no") have only one of these properties. The figure is from the experiments of Ian Dennis and his coworkers.*

of the set and that *m* properties suffice to allow membership, where *n* is larger than *m*. If *m* of the *n* properties permits membership, two members need not have any of the same properties in common. This partly defines what is called a polymorphous set.) Wittgenstein also considered several other intriguing ideas—that certain categories may have *degrees* of membership but no clear boundaries and that others may have members that are more central or prototypical than others.

Since Wittgenstein's time, psychologists have done a number of studies establishing evidence to support his ideas. Notable among them are Brent Berlin and Paul Kay, who showed that human color categories have degrees of membership and centrality; Roger Brown, who showed that children first name things at a level that is neither the most general nor the most specific; and Eleanor Rosch and her coworkers, whose studies were perhaps the most general and who developed the analysis of categorization as a wide-ranging research tool.

Rosch's work shows the existence of family resemblance, centrality, and prototypicality. Categories like "red" have fuzzy boundaries but nonetheless contain central members whose degree of membership on a scale from zero to one would be scored as one. These are graded categories. Categories like "bird" have sharp boundaries, but within these boundaries some birds are judged to be better examples than others—to be more "prototypical." Knowledge of category members is often organized around a basic level—a level which, in subjects tested by Rosch, reveals itself in terms of the ease of imagining and remembering membership, actions, and use. "Horse" would be a basic-level category and "quadruped" would not be.

If we accept family resemblance, it is not surprising that there is often no definite hierarchical relationship between superordinate and subordinate categories. Consistent with this is the fact that categories are heterogeneous in origin: the actual properties humans use to determine category membership are interactional and they depend on different biological, cultural, and environmental variables.

This empirical work was done on human subjects. Although aspects of it have been challenged from time to time, it has been generally confirmed. More recently, Lance Rips showed that neither similarity nor typicality fully account for the *degree* of category membership and that the reasoning involved in placing membership is often nondeductive. Lawrence Barsalou additionally showed that particular categories are not even represented by invariant concepts. There is great variability in the concepts representing a category; different individuals do not represent a category in the same way, and the same individual changes his or her views of category membership in different contexts. Consistent with these ideas, the seminal studies of Daniel Kahneman and Amos Tversky showed that human decision making and human category judgments often violate probability rules such as the conjunction rule, which states that a conjunction is never more probable than either of its constituents. In certain contexts some people actually do not doubt that the conjunction is more probable.

I have been concerned here with conceptual categories; perceptual categories were discussed in the body of this book. But what we have at hand is enough to state that if this work is correct, the objectivist model of the mind–world relationship is in hot water. For example, if categories that have centrality and prototypicality, such as those for color, exist in addition to classical categories, then the objectivist view is inadequate. Worse than that, the objectivist model cannot deal with the fact that *certain symbols do not match categories in the world.* Psychological work indicates, for example, that computer views of the mind cannot deal with categories of the mind and of language (see any poem) that fail to reflect categories in the world. Individuals understand events and categories in more than one way and sometimes the ways are inconsistent. As Mark Johnson points out, metaphor and metonymy are major modes of thought. Metaphor is the referral of the properties of one thing to those of another in a different domain. Metonymy allows a part or an aspect of a thing to stand for the whole thing. Both are incompatible with the objectivist view.

All of this spells trouble for mental representation. In order to function, mentalese requires an accurate unambiguous link to the external world. Often meaning cannot be so established and such a link cannot exist. Objects in the world are *not* labeled with dimensions or codes, and the way they are partitioned differs from person to person and from time to time. Indeed, the fixed semantics of mental representation cannot account for the occurrence of novelty in the world and, as will be apparent in my discussion of language, well-defined codes cannot exhaust the meaning of linguistic expressions. Meaning simply refuses to be bound by a fixed set of terms in a specific coding system. While representations must remain fixed, behavior changes in new contexts (unaccountably, in the objectivist view).

If this holds, the mind is not a mirror of nature. Thought is not the manipulation of abstract symbols whose semantics are justified by unambiguous reference to things in the world. Classical categories do not serve in most cases of conceptual categorization and they do not satisfactorily account for the actual assignment of categories by human beings. There is no unambiguous mapping between the world and our categorization of it. Objectivism fails.

Memory and Language

Another source of embarrassment for the computational or functionalist view of mind has to do with memory and its connection to the self and to language. I consider some special aspects of language in the next section, but for now note that the words of a natural language are not like the terms of a computer language. I pointed out in the last section that all computation is syntactical in nature and thus, unlike the use of words in a speech community, it cannot have meaning without a programmer. Moreover, functionalists often speak of propositional attitudes—beliefs, desires, wishes. But as Putnam has pointed out, beliefs and desires cannot be individuated without reference to an open-ended environment, one that is not characterized beforehand.

There is an additional problem: Human memory is not at all like computer memory. As we have already noted, internal codes and syntactic systems cannot adequately describe human memory. Memory has been variously and sometimes confusingly described as episodic (relating to past events in a life), semantic (relating to language), procedural (relating to motor acts), declarative (referring to statements), and so on. *Memory is a system property:* It differs depending on the structure of the system in which it is expressed. In biological systems, memory must not be confused with the mechanisms that are necessary for its establishment, such as synaptic change. Above all, biological memory is not a replica or a trace that is coded to represent its object.

In whatever form, human memory involves an apparently open-ended set of connections between subjects and a rich texture of previous knowledge that cannot be adequately represented by the impoverished language of computer science—"storage," "retrieval," "input," "output." To have memory, one must be able to repeat a performance, to assert, to relate matters and categories to one's own position in time and space. To do this, one must have a self, and a conscious self at that. Otherwise, one must postulate a little man to carry out retrieval (in computers, it is we, the programmers, who are the little men). How is the proposed functionalist model of an algorithmic mind to be accessed without an infinite regress of homunculi, one inside the other?

With the homunculus, we come to one of the great problems in considering the matter of the mind: the problem of accounting for intentionality itself. We have already shown that formal semantics cannot refer unambiguously to real states of affairs. But many of the causal aspects of our mental states depend on semantic contents. As Searle has stressed, semantic contents are meaningless without intentionality or the ability to refer to other states or objects. To carry out referral, a formal representation must become an intentional one. In human beings, this requires a consciousness and a self—a biologically based personal awareness, a first person. No theory of mind worth its salt can evade this issue, which is not only a matter of language but also a great biological problem. Let us pursue our quarry relentlessly, turning finally to some biological matters that cannot be reconciled with the functionalist picture of the mind.

The Lessons from Biology

A great revolution in thought came from Darwin's efforts to understand the origin of species. In his theory of natural selection, he gave the world its first example of population thinking. Population thinking, as Ernst Mayr puts it, considers variance to be real, that is, *not an error* (see figure 5–2). Natural selection acts on variation across individuals in a population. As Mayr has shown, species often arise as a result of the occurrence of sexual and geographic barriers to the propagation of variants or even by accident.

The species concept arising from this part of population thinking is central to all ideas of categorization. Species are not "natural kinds"; their definition is relative, they are not homogeneous, they have no prior necessary condition for their establishment, and they have no clear boundaries.

Thus, population thinking deals a death blow to typological thinking or essentialism, the notion that "essences" of species exist before particular organisms or exemplars do. Essentialism, most clearly formulated by Plato and reflected in most idealistic philosophies ever since, has a deep kinship to the notion of classical categories. But biology shows us that, even though taxonomy is possible for living creatures, essentialism is false. Given my previous remarks, it is also likely to be false in thinking about the mind.

Searle, Lakoff, Johnson, and others including myself have pointed out that thought is not transcendent but depends critically on the body and the brain. This position is exactly opposite to that of functionalism, which assumes that the realization of software is *independent* of hardware. According to those who reject functionalism, the mind is embodied. It is necessarily the case that certain dictates of the body must be followed by the mind. Gestalt perception is such a dictate; the categories of a gestalt (see, for example, figure 4–2) are not validated by a unique pattern *in the world*, and yet they are often incorrigible. Gestalts, mental images, bodily movements, and the organization of knowledge must all to some degree be the result of evolutionary and developmental constraints.

The syntax and semantics of natural languages are not just special cases of formal syntax and semantics, the models of which have structure but no meaning. In the biological view, symbols do *not* get assigned meanings by formal means; instead it is assumed that symbolic structures are meaningful *to begin with*. This is so because categories are determined by bodily structure and by adaptive use as a result of evolution and behavior. The symbols of cognition must match the conceptual apparatus contained in real brains. The bases for truth and knowledge come from this apparatus and have their earliest foundations in evolutionarily derived value systems. According to the purveyors of this view, including Lakoff, Johnson, Modell, and myself, when symbols fail to match the world directly, human beings use metaphor and metonymy to make connections, in addition to imagery and the perception of body schemes. Minds *create* aspects of reality through cultural and linguistic interaction. Like biology itself, this interaction depends on historical events. I deal directly with these matters in the next section, when I discuss language and its acquisition.

Besides embodiment, there is one more key issue—that of function. A deep insight into function was trenchantly expressed by Millikan in connection with minds, languages, and "other biological objects," as she calls them. Biological objects under evolution have functional properties that differ from, say, those of molecules. One does not speak of the "abnormal" function of a molecule as a chemical object. But a biological object has a proper function that depends on its evolutionary history. A heart has a proper function to pump blood. There is also what Millikan calls a "normal" explanation for the production of such an item in a species, and this accounts for the resemblance of this organ to "normal" hearts in that species. Hearts work well or not; badly functioning ones are abnormal. In contrast, organic chemicals do whatever they do, and *whatever* they do is part of their "working."

During evolution, functions that account for the proliferation of survivors are proper functions and associated with them is a "normal" explanation accounting for how they have *historically* managed to perform that function. The funny thing is that states and activities can have proper functions without performing them, and they can even have proper functions without contributing to further proper functions in accord with a "normal" explanation. This is because historical phenomena in selective systems can lead either to failure or to unexpected success.

Millikan regards psychology as a branch of biology, I think properly so. She claims that cognition gets its content from the identification of proper functions. This is an important claim. Each set of functions has a "normal" explanation that relates how the system manages to perform that function. Millikan's view of cognition allows it to be placed in the context of physiology (for example, that of the value systems I discussed in this book) and still provide grounds for a theory of beliefs and desires. Unlike the propositional attitudes of the functionalist, such a theory of intentionality does not differ dramatically from the uses and references of ordinary folk psychology (the way we usually characterize mental function in everyday life). In Millikan's view, the brain is thought of as a symbol manipulator and as a semantic engine. This is because beliefs and desires are "normally" manipulated in terms of significant (that is, *bodily* significant) differences between them and also in terms of differences in their proper functions. The appraisal of meaning and truth comes from this path, according to her analysis, not from the assignment of semantics by correspondences that are made by the "meaning rationalists," as she calls those with antithetical views.

The upshot of the arguments I have described is that the facts of biology force us to conclude that the mind is not transcendental. There is no God's-eye view of the world. Essentialism is not tenable and neither are functionalism, objectivism, or the form of "computational realism" that considers the mind to be a machine. Moreover, there is another fundamental source of difficulty that I described in the early chapters of this book and elsewhere: The variation in the structures and functions of the nervous system and the way in which the brain develops its anatomical connectivity by depending on correlations with events in the world are both incompatible with the functionalist view.

The vicious circles carved in the cognitive landscape are broken by the evidence underlying the foregoing analysis. But it is not enough to say that the mind is embodied to account for meaning and memory. The question is, How? And how, after explaining how, does this explanation account for the development of the self and of consciousness? This was the task I undertook in the body of the book. To accomplish it, I had to consider language in relation to higher-order consciousness. In addition, however, there are some technical matters specifically concerning language that must be placed in the context of the arguments pursued in this Postscript. Let us turn to them.

LANGUAGE: WHY THE FORMAL APPROACH FAILS

First, I want to consider how formal views of language are inconsistent with what I have already said about categories. Then I want to touch on some proposals for cognitive models and for grammars that are more in accord with what is known about categorization. My purpose is to contrast these two views, the formal and the cognitive, in an attempt to give the reader a glimpse of how different their premises are.

The study of language is vastly challenging and the field of linguistics proper is intricate in the extreme. I do not attempt to penetrate these studies here, for their full extent is beyond my expertise. Fortunately for our purposes, we need to know only a few guiding facts. I will describe them briefly and move on to my main point: that formal approaches to grammar fall under the same ax as felled the objectivist and the strict functionalist approaches to psychology.

To know a language is to be able to produce sounds or gestures conveying meaning and to understand them as they are produced by others. In general, the relation between form and meaning in a language is arbitrary. A striking feature of language is its creativity; a person competent in a language can make and understand completely new phrases and sentences. It is also striking that, most of the time, a person can tell the difference between proper and improper grammatical locutions.

Linguists attempt to construct theories of a speaker's grammar. In this sense, being grammatical means to conform to descriptive rules derived from use. In the broadest sense, however, grammar includes the study of the laws governing phonology (the sound system), morphology (in this context, word formation), and semantics (the system of meaning). All the laws together are said to make up a "universal grammar," a phrase that has been adopted from Noam Chomsky. According to his seminal proposals, all languages have a set of grammatical properties in common that constitute this universal grammar.

Chomsky has also suggested that inasmuch as language is unique to humans, and children's actual linguistic performance is underdetermined by their testable competence, there must be a "language acquisition device" that is innate to humans. It is important to note here that speaking is an *acquired* skill that develops from belonging to a speech community. A lot of categorizing must be done in order to speak. One must develop concepts or intentions, formulate expressions according to grammar and phonology, and articulate, comprehend, and monitor one's speech productions in exchanges with others.

To do this as an interlocutor requires a cooperative principle, as described by H. P. Grice. One must be informative at just the level required and no more; one must be brief, orderly, and unambiguous. One must be receptive to cues for turn taking. In addition, one must define the "here and now" or the "there and then." This so-called deixis locates both interlocutors and objects in space. One's communicative *intention* must also be expressed in appropriate speech acts. Speaking in general is a tactful as well as a tactical act.

Language acquisition in a speech community differs for children and adults. Moreover, to acquire a language and to use it are not necessarily the same thing. The study of linguistic knowledge, or psycholinguistics, and the study of the biological and neural bases of language, or neurolinguistics, must both be brought into play here. We confronted the differences between language acquisition and use in chapter 12. Recalling them here will keep us from confusing acquisition with rehearsed practice.

All of this is by way of introducing the problem of how thought and language are connected. A clear picture must be drawn of the relation between concept systems and language. Does the mastery of language depend on the existence of a rich and embodied concept system? Or is language mastery more or less autonomous, developing by means of a language acquisition device?

One of the most pervasive and influential approaches to these critical questions was pioneered by Chomsky. In his formal systems approach, the principal assumption is that the rules of syntax are independent of semantics. Language, in this view, is independent of the rest of cognition. I must take issue with this notion.

The set of rules formulated under the idea that a grammar is a formal system are essentially algorithmic. In such a system, no use is made of meaning. Chomsky's so-called generative grammar (figure P–7) assumes that syntax is independent of semantics and that the language faculty is independent of external cognitive capabilities. This definition of grammar is impervious to any attempt to disconfirm it by referring to facts about cognition in general. A language defined as a set of strings of uninterpreted symbols generated by production rules is like a computer language. To give the symbols semantic meaning, they must be mapped onto the real world or onto a language of thought or mentalese.

The previous discussion has prepared us for the conclusion that underlying this view is the objectivist position: categories are classical and semantics are generated by unambiguous assignment to entities in the world. This amounts to a definition of language and grammar. Under this definition, language falls afoul of all the difficulties faced by the objectivist view. The problem is not just that this view does not agree with the empirical facts concerning categorization. It also ignores the fact that language serves to convey the thoughts and feelings of individuals who already think independently of language.

The language acquisition device was proposed by Chomsky to answer the question of how a child who does not appear to understand many simple things can master the complexities of language. But a number of observations seem inconsistent with the Chomskian view. They concern thought and language acquisition in children as described, for example, in Margaret Donaldson's book *Children's Minds*. Donaldson points out that Chomsky directed the attention of his field toward studies of how a child acquires a knowledge of grammar. Consequently, linguists collected and interpreted data on what a child said in terms of a set of rules by which the child's utterances could have been generated. But in this enterprise, a good deal was ignored—often including what the child actually meant and what he or she understood.

As recounted by Donaldson, John Macnamara has proposed that children are able to learn language *because* they first make sense of situations

GENERATIVE GRAMMAR

PHRASE MARKERS: Symbolize the analysis of a sentence, "the girl was nice"

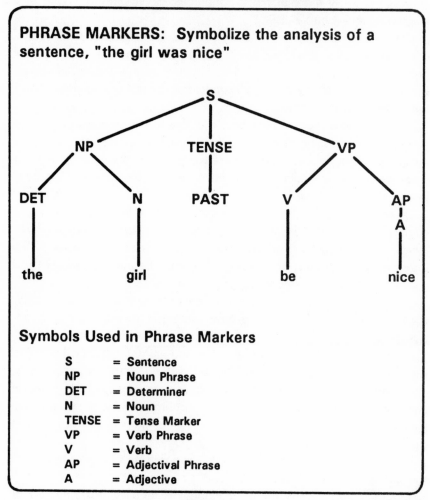

Symbols Used in Phrase Markers

S	= Sentence
NP	= Noun Phrase
DET	= Determiner
N	= Noun
TENSE	= Tense Marker
VP	= Verb Phrase
V	= Verb
AP	= Adjectival Phrase
A	= Adjective

FIGURE P–7

A typical tree in generative grammar, which is used to develop and analyze syntax. According to Chomsky, the rules of a universal grammar are assured by the presence in humans of an inborn language acquisition device that operates on such a syntax or on one of its modern exemplars. The relation to semantics is assured by the objectivist assumption (see figure P–5). This grammatical analysis has been superseded by Chomsky's more recent government binding theory, but the underlying assumptions remain unchanged.

involving human interactions. Children make sense of things *first* and, above all, they make sense of what people do. Donaldson's summary makes it clear that children can see things from another's point of view, not just their own. They reason deductively and carry out inference at age four or so, much more skillfully than had been previously supposed. It also seems that a child first makes sense of situations and of human intentions and *then* of what is said. This means that language is *not* independent of the rest of cognition. Therefore we need to account for language acquisition not only developmentally but also evolutionarily. I discussed this problem at length in chapter 12, where I considered how both conceptual *and* linguistic systems are embodied.

Before I turn to alternative ways of looking at language, I want to mention a prescient account by the novelist Walker Percy, whose collected essays on language were published in a book called *The Message in the Bottle*. My impression is that the attempt to understand language and meaning was at the center of his life and of his work. Percy was aware that generative or transformational grammar did not explain *language* and that it was merely a *formal* description of competence: No relationship is necessary between this collection of algorithms and what goes on in a person's head. He also understood that individual awareness is *symbolic* as well as intentional. Higher-order consciousness, as I have called it (see chapter 12), is a "knowing with" (con-sciousness). Percy faulted both behavioristic and semiotic approaches to language that do not pay attention to the *intersubjective* character of any linguistic act. He also faulted the philosophy of phenomenology for "leaving out the other guy." He insisted that all symbolic exchanges involving meaning show a tetradic relationship between symbol, object, and at least two humans. In a dense and resonant sentence, Percy put it thus: "The act of consciousness is the intending of the object as being what it is for both of us under the auspices of a symbol." He describes Helen Keller's rapture when she learned that water was "water" and her urgent desire to know then what other things "were." Language, as Percy put it, creates a world, not just an environment.

That world is loaded with intentionality, with projections, with feelings, with prejudice, and with affection. The story is told of two Jewish tourists who visited Israel for the first time. After an exhausting but enjoyable day in Tel Aviv, they decided to go to a nightclub. A comedian was on stage telling one-liners in Hebrew. After a few of these jokes, one of the tourists fell off his chair, laughing uncontrollably. His companion looked down and asked, "What are you laughing at? You don't even understand Hebrew." The man on the floor clutched his sides and said, "I trust these people."

Formal semantics cannot account for such richness. Well then, what can we do? One approach is to construct what is called a "cognitive grammar": an approach that begins with the facts of cognition rather than with formal analysis. An early pioneer was Ronald Langacker, whose book *Foundations of Cognitive Grammar* may be consulted for a history of this work and the principles guiding it. As in all nascent subjects, terminologies differ. Rather than use Langacker's, which is in a sense the "coining" terminology, I will, for the sake of convenience, discuss and follow the proposals of Lakoff, which are closely related to Langacker's and which provide examples closer to my own work in brain theory. Let us consider as an example this linguist's attempt to provide a model of cognition adequate to the available facts of categorization and to build a semantics based on the idea that meaning is embodied.

Cognitive Models and Cognitive Semantics: Returning to Biology

Lakoff has approached the subject of grammar and semantics in a way that appears to be more in accord with the biological and psychological facts than are generative grammars. He starts from actual data on categorization and proposes that meaning results from the intrinsic workings of the body and the brain. He suggests that individual humans construct cognitive models that reflect concepts concerned with the interactions between the body–brain and the environment. It is this conceptual embodiment, he claims, that leads to the formulation of basic-level categories of the kind described by Rosch.

Cognitive models are *created* by human beings, and in this sense they are idealized—that is, they are abstractions. But they depend on the formation of images as a result of sensory experience and they also depend on kinesthetic experience—the relation of the body to space. Lakoff suggests that the exercise of these functions leads to various image and kinesthetic schemas. Schemas have properties that are reflected later in the use of metaphor and metonyms. Recall that a metaphor is the referral or mapping of one thing to another in a different domain, while a metonym is the use of some part or aspect of a thing to stand for the thing itself. Lakoff's example of a metaphor is "Anger is a dangerous animal." His example of a metonym is "The ham sandwich left without paying."

The important thing to grasp is that idealized cognitive models involve conceptual embodiment and that conceptual embodiment occurs through bodily activities *prior to language.* Conceptual embodiment is used in categorization and it allows for the heterogeneity and complexity of real human categorization. Categories of mind correspond accordingly to elements in cognitive models. Some of these models have different degrees of membership. Others include classical categories and are formed according to singly necessary and jointly sufficient conditions (note that there is no contradiction here, provided that not *all* the models are classical!). Some models are metonymic. But the most complex cognitive models correspond to what Lakoff calls radial categories. These consist of many models linked around a center. Although noncentral models (and categories) cannot be predicted by a knowledge of the central category, they do have a relationship to the center; they are said by Lakoff to be "motivated" by it.

Such properties permit degrees of membership, degrees of relationship to the central model, family resemblances, nonhierarchical relationships with basic categories dominating, and prototype effects. Prototype effects are not fundamental but arise from many sources—"scalar," "classical," "metonymic," and "radial."

With this background, Lakoff attempts to mount a structure for cognitive *semantics* (figure P–8). Notice first that meaning is already based in embodiment by means of image schemas, kinesthetic schemas, metonyms, and the categorical relations that underlie metaphor. But this is not enough: Language is supposed to be characterized by *symbolic* models. These are models that pair linguistic information with the cognitive models that themselves make up a *preexisting* conceptual system. Inasmuch as preexisting conceptual models are already embodied through their link to bodily and social experience, this link is not an arbitrary one. In contrast, the attribution of such a linkage to generative grammar in terms of mental representations *is* arbitrary; it is made from on high by the grammarian.

In this view of cognitive semantics, linguistic categories naturally show strong structural resemblances to their underlying cognitive models. Language makes use of general cognitive mechanisms to construct propositional models, image schematic models, metaphoric models, and metonymic models. As we have said, metaphoric models involve a mapping from a structure in one domain to a corresponding structure in another domain. This mapping involves either propositional or image schemas. Metonymic models use these schemas and metaphor to map a function from one element of a model to another (for example, a part–whole relationship).

In *Women, Fire, and Dangerous Things: What Categories Reveal About the Mind,* Lakoff uses the work of his colleague Johnson (*The Body in the Mind: The Bodily Basis of Meaning, Imagination, and Reason*) to construct a series

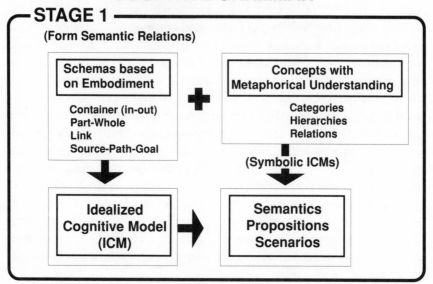

COGNITIVE GRAMMAR

STAGE 1

(Form Semantic Relations)

Schemas based on Embodiment

Container (in-out)
Part-Whole
Link
Source-Path-Goal

+

Concepts with Metaphorical Understanding

Categories
Hierarchies
Relations

(Symbolic ICMs)

Idealized Cognitive Model (ICM)

Semantics Propositions Scenarios

STAGE 2

(Build Syntax from Semantics and the Categorization of Sentence and Phrase Relations)

Syntactic ICM

Constructed using same schemas:

Hierarchical Syntactic Structure   - Part-Whole Schema
Grammatical Relations   - Link Schemas
Syntactic Categories   - Container Schema

FIGURE P–8

An example of processes in a cognitive grammar according to Lakoff. In contrast to generative grammar (see figure P–7), rules are acquired through linguistic experience and meanings arise because concepts are embodied. This kind of grammar has not yet been shown fully to have the analytic power of mainline generative attempts such as the lexical functional grammar of Bresnan. But it does give a scheme for the relation of meaning (through embodiment) to categorization and sentence structure. The "stages" are not necessarily sequential in time and they overlap. Obviously they have the least overlap during early language acquisition. ICM = idealized cognitive model.

of schemas, based on embodied concepts, that provide a basis for linguistic meaning. These include *container* schemas (defining a boundary, or "in and out"), a *part–whole* schema, a *link* schema (one thing connected to another, as by a string), a *center–periphery* schema (as in body center versus arms and legs), and a *source–path–goal* schema (starting point, directional path, midpoint) including *up–down* and *front–back* schemas. He then goes on to show that metaphors are motivated by the structuring of experience resulting in schemas. The *source–path–goal* schema, for example, emerges from our bodily functioning, pervades our experience, is well structured, and is well understood. The source domain as well as the target domain of any metaphor based on it will be correlated experientially in terms of this schema. The prior basic-level and image schematic concepts are directly meaningful and provide the basis for the schema. They also provide starting points for the rules of semantic composition that form more complex concepts from simpler ones.

These ideas are encapsulated in Lakoff's "spatialization of form hypothesis." According to this hypothesis, categories are understood in terms of *container* schemas, hierarchical structure is understood in terms of *part–whole* and *up–down* schemas, relational structure is understood in terms of *link* schemas, the radial structure of categories is understood in terms of *center–periphery* schemas, foreground–background structure is understood in terms of *front–back* schemas, and linear quantity scales are understood in terms of *up–down* and *linear order* schemas. All involve a metaphorical mapping from physical (or spatial) structures to conceptual structures.

But where, specifically, does language come in? With *idealized cognitive models*, some of which are structures consisting of symbols. These models are of five types: image schematic, metaphoric, metonymic, propositional, and symbolic. Of these, the ones that lead to linguistic function are the propositional and symbolic idealized cognitive models.

A propositional idealized cognitive model does not use metaphor, metonymy, or mental imagery. Instead, it uses basic-level concepts—entities, actions, states, and properties. Simple propositions follow the part–whole schema: the proposition is the whole, of which the predicate is one part and the arguments (agent, patient, experiencer, instrument, location, and so on) are the other. Semantic relations are built from link schemas, and complex propositions are then formed from simple propositions by modification, quantification, conjunction, negation, and so on. Moreover, scenarios can be built of an initial state, a sequence of events, and a final state structured by a *source–path–goal* schema.

When linguistic elements are associated with *conceptual* idealized cognitive models, these become *symbolic* idealized cognitive models. They can then be characterized in terms of the morphemes and words of particular languages. A noun, for example, is a radial category (central categories are people, places, things; noncentral categories are abstract nouns like "strength"). A *verb* is also a radial category (central categories are basic-level physical actions like run, hit, give). Remaining members of these categories are motivated by relations to these central members. The relation to semantics is obvious.

And what of syntax itself? Lakoff claims that the principles he discusses allow us to provide a semantic basis for syntactic categories. According to his theory, hierarchical syntactic structure (see figure P–7 for an example) is itself characterized by *part–whole* schemas, head and modifier structures are characterized by *center–periphery* schemas, grammatical relations are characterized by *link* schemas, and syntactic categories are characterized by *container* schemas. Notice what is happening here: *Grammatical constructions are themselves idealized cognitive models.* Thus, semantic pairing with syntax is a direct pairing of an idealized cognitive model for syntax with a *prior* idealized cognitive model for semantics or meaning. Regularities in the structure of grammar and in a lexicon can be described in terms of radial categories, and words with multiple meanings can be explained in these terms.

In this view, and in Langacker's, language is based on cognition—that is, on cognitive models that can be understood in terms of bodily functioning. This cognitive base is constrained by the nature of physical reality and also depends on imagination and social interactions. Meaning derives from embodiment and function, understanding arises when concepts are meaningful in this sense, and truth is considered to arise when the understanding of a statement fits one's understanding of a situation closely enough for one's own purposes. (Notice the pragmatism!) Thus, there is no absolute truth or God's-eye view. Our view of what exists (metaphysics) is not independent of how we know it (epistemology). As Lakoff puts it, "Truth is a bootstrapping operation, grounded in direct links to preconceptually and distinctly structured experience and the concepts that accord with that experience." This fits with my proposals related to qualified realism in chapter 15.

Knowledge, like truth, is a radial concept. It depends on our understanding, on basic-level concepts, and also on socially accepted understanding. It is secure to the extent that human understanding can be secure, but it is always subject to revision. Objectivity is not absolute but depends on looking at a situation from as many points of view as possible and by distinguishing basic-level concepts and image schematic concepts from concepts that are only indirectly meaningful.

Obviously the example provided by Lakoff's cognitive grammar (figure P–8) is radically different from the more widely accepted generative grammars (see figure P–7). It differs in philosophy, style, and methodology. It is in closer accord with the biological bases of brain and bodily function and with the psychological data on categorization. It avoids the category mistakes of the "language of thought" proposal and the objectivist error inherent in generative grammar. It is an imaginative and important proposal. But in proposing embodiment as the origin of meaning, it does not show *how* this might come to pass. Nor does it show how symbolic idealized cognitive models of language arise as a result of the mechanisms of perceptual and conceptual categorization. For these tasks, one needs a general biological theory of brain function and a theory of consciousness, both based on the facts of evolution and development. That is what I attempted to construct in my trilogy and to review in this volume.

It may be useful to comment on the relation between Lakoff's cognitive grammar and the theory of speech acquisition described in chapter 12. Cognitive grammar is based on the notion of embodiment, but it does not specify how such embodiment takes place. Instead it searches for signs of radial categories, metaphor, and metonymy as guiding structures for speech. And, similarly, it uses categorization to account for the emergence of syntactical relationships. In all these respects, it is compatible with the epigenetic theory presented in chapter 12. This theory clarifies the issues related to evolution and to the acquisition of speech in a way that Lakoff's theory, lacking a description of mechanisms of embodiment, cannot. Indeed, the epigenetic theory provides additional grounds for taking aspects of an extensive *structural* generative theory of grammar such as Bresnan's (which stresses lexicon) and linking them to a *categorically based* theory such as Lakoff's. Langacker's treatment, while rejecting the generative aspects of Bresnan's approach, resembles it in stressing the importance of lexicon. A complete understanding of grammatical formulations requires an analysis of the brain mechanisms for concept formation, value-category formation, connection to the phenotype, and connection to the mechanisms of consciousness. It also requires exploration of the grammars of particular languages in the terms described by Langacker, Bresnan, and others. A rich field of study might be opened up by exploring how a theory of embodiment such as the one I have presented might bridge and relate these important but different approaches to linguistic theory.

Lakoff's *Women, Fire, and Dangerous Things* came out at about the same time as my book *Neural Darwinism*, which attempted to provide a basis for a global brain theory. I know that I was unaware of his book and surmise that he was unaware of mine. The central problem confronted by *Neural*

Darwinism was perceptual categorization. In a subsequent work, *The Remembered Present: A Biological Theory of Consciousness,* I extended the brain theory to perceptual experience, concept formation, and language. In retrospect, it appears that these two books nicely complement Langacker's, Lakoff's, and Johnson's work, providing an essential biological underpinning for many of their proposals concerning the importance of embodiment to grammar and cognition. But neither their work nor mine denies the significance of the efforts of other linguists to understand syntactic structure. The importance of their efforts and of the efforts of cognitive psychologists is very great. But without biology, they remain insufficient and even, at times, in error. This is what I have attempted to show in this Postscript.

For those who have read both the text and the Postscript, I hope the challenge has been made sufficiently clear. We must incorporate biology into our theories of knowledge and language. To accomplish this we must develop what I have called a biologically based epistemology—an account of how we know and how we are aware in light of the facts of evolution and developmental biology. A fuller realization of this goal will expand our scientific horizons. And through its connections to what makes us uniquely human, a biologically based epistemology will enrich our lives.

Selected Readings

I have put together the following references for those readers who want to obtain additional background or follow up an idea. Where I have mentioned some recent original work in the text, I have attempted to cite a paper and hold at least to the minimal requirements of scholarship.

For extensive bibliographies, the reader may consult those in my trilogy on morphology and mind: *Topobiology, Neural Darwinism,* and *The Remembered Present* (all published by BasicBooks and listed below).

The comments following the given references may be useful to some readers. They reflect my personal opinions but are not at all complete.

CHAPTER 1

JAMES, W. *The Principles of Psychology,* 1890. Reprint. New York: Dover, 1950.
A monumental work by one of the founders of experimental psychology. Contains penetrating descriptions and analyses, as well as strong personal opinions.

JAMES, W. "Does Consciousness Exist?" In *The Writings of William James,* ed. J. J. McDermott. Chicago: University of Chicago Press, 1977, 169–83.
A seminal work that makes the case that consciousness is a process, not a thing or substance.

FLANAGAN, O. J., JR. *The Science of the Mind,* 2nd ed. Cambridge, Mass.: MIT Press, 1991.
A nice survey of the thoughts and work of modern psychologists, with a balanced assessment of their present standing.

BRENTANO, F. *Psychology from an Empirical Standpoint*, ed. O. Kraus and L. L. McAlister, trans. A. C. Rancurello et al. Highlands, N.J.: Humanities, 1973.
The key work of the psychologist, philosopher, and ex-priest who emphasized the importance of intentionality. A professor in Vienna, he influenced Freud, who attended his lectures.

GREGORY, R. L. *Mind in Science*. Cambridge: Cambridge University Press, 1981.
An historical account of the uneasy relationship between scientific methodology and the matter of the mind. Discursive, inconclusive, but rich in suggestions.

GRIFFIN, D. R. *Animal Thinking*. Cambridge, Mass.: Harvard University Press, 1984.
A spirited case for animal awareness made by a noted ethologist. In other works he has gone so far as to suggest that bees are conscious. I believe the case has scarcely been made (and indeed is highly unlikely), but Griffin's way-out position is usefully provocative.

PREMACK, D., and A. J. PREMACK. *The Mind of an Ape*. New York: Norton, 1983.
This is a splendid and clear account of the mental capabilities of chimpanzees. Current, expert, and stimulating.

MCCULLOCH, W. S. *Embodiments of Mind*. Cambridge, Mass.: MIT Press, 1989.
Essays by a brilliant forerunner of modern neurobiology, reprinted. Useful and imaginative.

BLAKEMORE, C., and S. GREENFIELD, eds. *Mindwaves: Thoughts on Intelligence, Identity, and Consciousness*. New York: Blackwell, 1987.
A collection of papers on these subjects by neuroscientists and philosophers. Displays the major issues, the confusions, and the different positions of various practitioners.

GREGORY, R. L., ed. *The Oxford Companion to the Mind*. Oxford: Oxford University Press, 1987.
A small encyclopedia with articles on many issues by a variety of experts. Spotty, but useful and lots of fun for the intellectually curious. Great for browsing.

CHAPTER 2

WHITEHEAD, A. N. *Science and the Modern World*. New York: Macmillan, 1925.
The classic account by a logician, historian of science, and metaphysician. Puts the relation between the scientific observer and the subjectively reflective individual in a rich historical perspective.

GALILEI, G. "The Assayer," 1623, trans. S. Drake. In *Discoveries and Opinions of Galileo*. New York: Doubleday, 1957. DRAKE, S. *Galileo*. Oxford: Oxford University Press, 1980.
In both books, the "founder" is on display and his thoughts summarized by one of his most devoted historians. Those who feel that scientists today are too

concerned with priority should read Galileo's complaints in the beginning of "The Assayer." Incidentally, Galileo understood the nature of secondary qualities (color, warmth, and so forth) almost a century before Locke.

DESCARTES, R. *Meditations* and *Passions of the Soul.* In *The Philosophical Works of Descartes,* vols. 1 and 2, ed. E. Haldane and G. Ross. Cambridge: Cambridge University Press, 1978.

If Galileo is the founder of modern science, Descartes is the founder of modern philosophy. His thoughts are proof that genius, even genius leading to wrong conclusions, can be of continuing major significance. We still wrestle with the questions that Descartes posed.

CHAPTER 3

SHEPHERD, G. *Neurobiology.* New York: Oxford University Press, 1983.

A standard elementary account of modern findings. Many others exist, but this has most of the basics, with some discussions at the level of principles.

LURIA, A. R. *The Working Brain: An Introduction to Neuropsychology.* New York: Basic Books, 1973.

The late, great clinician and neurologist from the Soviet Union gives a clear account of what happens when parts of the brain are disturbed. In following the course of his descriptions, a newcomer receives a larger picture of the functions of the brain proper than can be culled from a book like Shepherd's.

DIDEROT, D. *Le Rêve de d'Alembert,* 1769. Reprint. New York: Penguin, 1966.

The little classic quoted in the text. The original is in St. Petersburg; Diderot was an adviser to Catherine the Great. My understanding is that his collaborator on the Encyclopedia, d'Alembert, was none too pleased to have his relationship with Mlle de l'Espinasse so openly on display.

CHANGEUX, J.-P. *Neuronal Man: The Biology of Mind.* New York: Oxford University Press, 1986.

A popular account by a neurobiologist who believes, as I do, that the brain is a selectional system. Contains quick surveys, historical matters, and a short account of connections between cognitive psychology and neurophysiology.

CHAPTER 4

For a history of philosophy at the elementary level, see B. RUSSELL, *A History of Western Philosophy.* New York: Simon and Schuster, 1945. This is clear, prejudiced, and stimulating. For more modern developments astringently done, see A. J. AYER, *Philosophy in the Twentieth Century.* East Hanover, N.J.: Vintage, 1984. A good (if rather technical) set of historical accounts of modern psychology may be found in E. Hearst, ed. *The First Century of Experimental Psychology.* Hillsdale, N.J.: Lawrence Erlbaum Associates, 1979.

Kanizsa, G. *Organization in Vision: Essays on Gestalt Perception.* New York: Praeger, 1979.
Wonderfully revealing account of the world of so-called visual illusions. Masterful and a minor classic, with marvelous illustrations.

CHAPTER 5

Darwin, C. *On the Origin of Species by Means of Natural Selection or the Preservation of Favoured Races in the Struggle for Life.* London: Murray, 1859.
The masterpiece that provided the foundation for modern biology.

Barrett, P. H., P. J. Gautrey, S. Herbert, D. Kohn, and S. Smith, eds. *Charles Darwin's Notebooks, 1836–1844: Geology, Transmutation of Species, Metaphysical Enquiries.* Ithaca: Cornell University Press, 1987.
Views into the mind of a great scientist and thinker.

Romanes, G. J. *Mental Evolution in Animals.* New York: Appleton, 1884. *Mental Evolution in Man.* New York: Appleton, 1889.
The thoughts of Darwin's contemporary. Good examples of how a great provocative theory takes root in widespread territories.

Mayr, E. *The Growth of Biological Thought: Diversity, Evolution, and Inheritance.* Cambridge, Mass.: Harvard University Press, 1982.
A modern masterpiece by a great evolutionist. One of the best accounts of Darwin, Darwinism, and the "subtheories" that make up a complex theory like the modern theory of evolution.

Richards, R. J. *Darwin and the Emergence of Evolutionary Theories of Mind and Behavior.* Chicago: University of Chicago Press, 1987.
A very comprehensive exposition, rich in scholarship. The best up-to-date account of this subject.

CHAPTER 6

Thompson, D. W. *On Growth and Form.* Cambridge: Cambridge University Press, 1942.
One of the great classics on the subject of animal form, by a talented nonbeliever in Darwin. While its direct examples are not very relevant to the nervous system, it is nonetheless fascinating.

Edelman, G. M. *Topobiology: An Introduction to Molecular Embryology.* New York: Basic Books, 1988.
A more extended and fundamental account of the subject of this chapter. The last chapter of *Topobiology,* which is by virtue of its subject matter the first

volume of the trilogy on morphology and mind (although not the first to be published), describes the connection between topobiology and selectionist theories of the brain.

CHAPTER 8

BURNET, F. M. *The Clonal Selection Theory of Acquired Immunity.* Nashville: Vanderbilt University Press, 1959.
This is the original extended account of Burnet's selectionist views. For the evolutionary background, see Mayr's book, cited above.

CHAPTER 9

EDELMAN, G. M. *Neural Darwinism: The Theory of Neuronal Group Selection.* New York: Basic Books, 1987.
This book lays out the theory of neuronal group selection *in extenso* with its defenses well up. Much more scholarly than the present account. The original skeletal exposition of the theory may be found in G. M. Edelman and V. B. Mountcastle, *The Mindful Brain,* Cambridge, Mass.: MIT Press, 1978.
BARLOW, H. B. "Neuroscience: A New Era?" *Nature* 331, no. 18 (February 1988): 571. CRICK, F. "Neural Edelmanism." *Trends in Neurosciences* 12, no. 7 (July 1989): 240–48. PURVES, D. *Body and Brain: A Trophic Theory of Neural Connections.* Cambridge, Mass.: Harvard University Press, 1988.
These three authors attack aspects of the theory of neuronal group selection. The abbreviated counterattacks are presented in this chapter. Vive le sport!
MICHOD, R. E. "Darwinian Selection in the Brain." *Evolution* 43, no. 3 (1989): 694–96.
A favorable review agreeing that the theory of neuronal group selection is a selectionist account in the spirit of population thinking. See also his reply to Crick's criticism of the TNGS, along with the accompanying reply of G. N. Reeke, Jr., in *Trends in Neurosciences* 13, no. 1 (1990): 11–14.
For more information on the work of Eckhorn and his colleagues and Gray and his colleagues, see the references in Sporns, O., J. A. Gally, G. N. Reeke, Jr., and G. M. Edelman, "Reentrant Signaling Among Simulated Neuronal Groups Leads to Coherency in Their Oscillatory Activity." *Proceedings of the National Academy of Science* 86 (1989): 7265–69.

CHAPTERS 10 THROUGH 13

EDELMAN, G. M. *The Remembered Present: A Biological Theory of Consciousness.* New York: Basic Books, 1989.
The last of the trilogy on morphology and mind. An attempt to provide a

principled scientific account of the bases of consciousness. Although it is the last volume of the trilogy, I am told by some that it is best read first.

MARCEL, A. J., and E. BISIACH, eds. *Consciousness in Contemporary Science.* Oxford: Clarendon, 1988.

A collection of valuable papers on the subject, ranging over a wide area.

BARTLETT, F. C. *Remembering: A Study in Experimental and Social Psychology.* Cambridge: Cambridge University Press, 1964. *Thinking: A Study of Humans.* New York: Basic Books, 1958.

Two classics, especially the first, which gives a profound analysis of the act of remembering.

HEBB, D. O. *The Organization of Behavior: A Neuropsychological Theory.* New York: Wiley, 1949.

HEBB, D. O. *Essay on Mind.* Hillsdale, N.J.: Lawrence Erlbaum Associates, 1980.

The first is one of the earliest attempts to account for psychological phenomena in terms of neuronal and cellular interactions. The second contains the later reflections of a modern master.

FREUD, S. "Project for a Scientific Psychology." In *The Standard Edition of the Complete Psychological Works of Sigmund Freud,* vol. 1, ed. J. Strachey. London: Hogarth, 1976, 283–411. *Introductory Lectures on Psychoanalysis.* New York: Liveright, 1919. *New Introductory Lectures on Psychoanalysis.* New York: Norton, 1933. *On Dreams,* ed. J. Strachey. Reprint. New York: Norton, 1963.

"Project for a Scientific Psychology" elated Freud and then repelled him. It was saved for posterity by his friend, Marie Bonaparte. The rest of these works represent the best kernels of the master's work at a popular level. The chef d'oeuvre, *On Dreams,* was the work he considered to be his greatest.

ERDELYI, M. H. *Psychoanalysis: Freud's Cognitive Psychology.* New York: Freeman, 1985.

An excellent account of key Freudian concepts, particularly Freud's ideas on memory and the unconscious.

BROUWER, L. E. J. "Consciousness, Philosophy, and Mathematics." In *Proceedings of the Tenth International Congress of Philosophy,* vol. 2, ed. E. W. Beth, H. J. Pos, and J. H. A. Hollak. Amsterdam: North Holland, 1949, 1235–49.

A remarkable piece of imagination by a topologist and philosopher of mathematics. Post-Kantian, hard to grasp, but very stimulating.

HILGARD, E. R. *Divided Consciousness: Multiple Controls in Human Thought and Action.* Expanded edition. New York: Wiley, 1977.

A different view, decidedly post-Freudian. Full of fascinating examples and discussions of split and multiple consciousness and hypnotic phenomena. For a startling exploration of visual Gestalt phenomena, see the marvelous book by Gaetano Kanizsa that is listed in the readings for chapter 4.

STADDON, J. E. R. *Adaptive Behavior and Learning.* Cambridge: Cambridge University Press, 1983.

A good account of learning seen from a wide biological base.

ALEXANDER, R. D. "Evolution of the Human Psyche." In *The Human Revolution,* eds. P. Mellars and C. Stringer. Princeton, N.J.: Princeton University Press, 1989.

Self-deceit as an adaptive phenomenon in the deceiving of others during the struggle for survival—an unlikely but fascinating hypothesis.

CHAPTER 14

ROTA, G. C. "Mathematics and Philosophy: The Story of a Misunderstanding." *Review of Metaphysics* 44 (December 1990): 259–71.

A pungently phrased and thoughtful article by a distinguished mathematician on the distrust one should have of excessive reliance on axiomatics: "The snobbish symbol dropping one finds nowadays in philosophical papers raises eyebrows among mathematicians. It is as if you were at the grocery store and you watched someone trying to pay his bill with Monopoly money."

CHAPTERS 15 AND 16

EDWARDS, P., ed. *The Encyclopedia of Philosophy*, vols. 1–4. New York: The Free Press, 1973. DANTO, A. *Connections to the World: The Basic Concepts of Philosophy*. New York: Harper & Row, 1989. RUSSELL, B. *The Problems of Philosophy*. 1912. Reprint. Oxford: Oxford University Press, 1959.

A reference work and two lucid introductions, one new, one old.

BECHTEL, W. *Philosophy of Mind: An Overview for Cognitive Science*. Hillsdale, N.J.: Lawrence Erlbaum Associates, 1988. CHURCHLAND, P. M. *Matter and Consciousness*. Cambridge: MIT Press, 1984.

Two short introductions to the philosophy of mind.

QUINE, W. V. *Quiddities: An Intermittently Philosophical Dictionary*. Cambridge: Harvard University/Belknap Press, 1987.

Amusing idiosyncratic notes by an outstanding American philosopher and logician.

WITTGENSTEIN, L. *Philosophical Investigations*. The English text translation by G. E. M. Anscombe. New York: Macmillan, 1953.

The posthumously published revisionist views of one of the most interesting philosophic minds of this century. Bears on the issue of categorization discussed throughout the present work.

WHITEHEAD, A. N. *Modes of Thought*. New York: The Free Press, 1938.

Metaphysical reflections, *tout court*, by a modern thinker now temporarily fallen from grace in most university circles. Well worthwhile for its imaginative and suggestive insights.

NAGEL, T. "What Is It Like to Be a Bat?" *Philosophical Review* 83 (1974): 435–50. *The View From Nowhere*. New York: Cambridge University Press, 1986.

Penetrating and clear analyses of the dilemmas of epistemology and metaphysics.

RYLE, G. *The Concept of Mind*. Chicago: University of Chicago Press, 1949.

A searing critique of category errors in the philosophy of mind by the originator of the phrase "the ghost in the machine."

RUSSELL, B. *A History of Western Philosophy.* New York: Simon & Schuster, 1945.
Already mentioned. Lucid, *sui generis;* an account by one of the pioneers of mathematical logic and one of the most courageously opinionated of modern philosophers.

AYER, A. J. *The Problem of Knowledge.* Middlesex, N.J.: Penguin, 1956.
A view of the whole issue in epistemology from a former logical empiricist.

PIAGET, J. *Biology and Knowledge: An Essay on the Relations Between Organic Relations and Cognitive Processes.* Chicago: University of Chicago Press, 1971.
The views, presented here for contrast, of a great developmental psychologist. Not only idiosyncratic and original but revealing of the gulf between the attitudes of scientists and philosophers. Somewhat metaphorical in its comparisons of embryology and psychology.

DAVIS, P. J., and R. HERSH. *Descartes's Dream: The World According to Mathematics.* Boston: Houghton Mifflin, 1986.
A beautiful account by two mathematicians of the nature and limits of mathematics. Unsympathetic to the Platonic view of mathematics.

MORGAN, M. J. *Molyneux's Question: Vision, Touch, and the Philosophy of Perception.* Cambridge: Cambridge University Press, 1977.
An elegant essay on some historical aspects of the psychology of spatial perception. Never mind the view of a bat—what would happen if you were always blind and then suddenly regained your sight? Would your "touch space" and "visual space" correspond?

HULL, J. M. *Touching the Rock: An Experience of Blindness.* New York: Pantheon, 1990.
A moving account of how an individual's consciousness is altered by the loss of vision.

BOYD, R., and P. J. RICHERSON. *Culture and the Evolutionary Process.* Chicago: University of Chicago Press, 1985.
A remarkably balanced account of how human social behavior and evolution may interact. One of the best forays into this dangerous thicket.

BARROW, J. D., and F. J. TIPLER. *The Anthropic Cosmological Principle.* Oxford: Oxford University Press, 1988.
Source of my quote on wastepaper baskets. Perhaps the relatively large number of references on philosophy contained in this section proves the contention of that quote (see page 159).

FLEW, A. *An Introduction to Western Philosophy: Ideas and Arguments from Plato to Popper.* New York: Thames and Hudson, 1989.
Contains a good discussion of the idea of the soul.

See also the references to chapter 4.

CHAPTER 17

ARENDT, H. *The Life of the Mind,* vol. 1, *Thinking;* vol. 2, *Willing.* San Diego: Harcourt Brace Jovanovich, 1978.

A philosophical résumé. It is revealing to contrast its views with those of experimentalists like Bartlett (for references to Bartlett, see the selected readings for chapters 10–13).

LANGER, S. K. *Mind: An Essay on Human Feeling*, vols. 1–3. Baltimore: Johns Hopkins University Press, 1967, 1972, 1973.

A masterful survey by a philosopher, historian of ideas, and scholar of artistic symbolism. Poignantly cut short in volume 3 by the author's advancing blindness.

MANDLER, G. *Mind and Body: Psychology of Emotion and Stress*. New York: Norton, 1984. SOLOMON, R. C. 1978. *The Passions*. Notre Dame: University of Notre Dame Press, 1983. DE SOUSA, R. *The Rationality of Emotion*. Cambridge: MIT Press, 1987.

Three accounts of emotion, the first scientific, the last two philosophical. Together, they bring out the extraordinarily complex, multilevel nature of emotions.

CHAPTER 18

WILLIAMS, M. *Brain Damage, Behavior, and the Mind*. New York: Wiley, 1979. KOLB, B., and I. Q. WHISHAW. *Fundamentals of Human Neuropsychology*, 3rd ed. San Francisco: Freeman, 1990. MCCARTHY, R. A., and E. K. WARRINGTON. *Cognitive Neuropsychology: A Clinical Introduction*. New York: Academic, 1990.

Three books on the effects of brain damage (see also A. Luria, *The Working Brain*, in the references to chapter 1).

KAPLAN, H. I., and B. SADOCK. *Comprehensive Textbook of Psychiatry/IV*, vols. 1–2. Baltimore: Williams & Wilkins, 1989.

A large psychiatry text. For the brave.

MODELL, A. H. *Other Times, Other Realities: Towards a Theory of Psychoanalytic Treatment*. Cambridge: Harvard University Press, 1990. HUNDERT, E. M. *Philosophy, Psychology, and Neuroscience: Three Approaches to the Mind*. Oxford: Clarendon, 1989.

Two psychiatrists apply the theory of neuronal group selection to aspects of their subject.

SCHACTER, D. L., M. P. MCANDREWS, and M. MOSCOVITCH. "Access to Consciousness: Dissociations Between Implicit and Explicit Knowledge in Neuropsychological Syndromes." In *Thought Without Language*, ed. L. Weiskrantz. Oxford: Clarendon, 1988, 242–78. BISIACH, E. "Language Without Thought." In *Thought Without Language*, ed. L. Weiskrantz, Oxford: Clarendon, 1988, 465–91.

Two revealing articles on some dissociative syndromes of consciousness.

SACKS, O. *The Man Who Mistook His Wife for a Hat and Other Clinical Tales*. New York: Harper & Row, 1987.

A fascinating set of accounts by a humane mind, a clinician, and a magnificent storyteller.

CHAPTER 19

Reeke, G. N., Jr., and G. M. Edelman. "Real Brains and Artificial Intelligence." *Daedalus* 117, no. 1, (Winter 1988); 143–73. See also Reeke, G. N., Jr., L. H. Finkel, O. Sporns, and G. M. Edelman. "Synthetic Neural Modeling: A Multilevel Approach to the Analysis of Brain Complexity." In *Signal and Sense: Local and Global Order in Perceptual Maps*, eds. G. M. Edelman, W. E. Gall, and W. M. Cowan, New York: Wiley-Liss, 1990: 607–707. Edelman, G. M., and G. N. Reeke, Jr. "Is It Possible to Construct a Perception Machine?" *Proceeds of the American Philosophical Society* 134, no. 1 (1990): 36–73.

These articles reflect the philosophy underlying a major research program at The Neurosciences Institute. The second is quite technical and extensive. The third offers some reflections on the relation of this work to other work in the field. No article on NOMAD—the "real thing"—is yet in print, but undoubtedly one will be published by the time this book appears.

CHAPTER 20

Adair, R. K. *The Great Design: Particles, Fields, and Creation.* New York: Oxford University Press, 1987.

A beautiful summary of modern physics with a smidgeon of cosmology. Technical, but worth the effort.

Zee, A. *Fearful Symmetry: The Search for Beauty in Modern Physics.* New York: Macmillan, 1986. Weyl, H. *Symmetry.* Princeton: Princeton University Press, 1952. Tarasov, L. *This Amazingly Symmetrical World.* Moscow: Mir, 1986.

Three books on the significance of symmetry in the field of physics and elsewhere. Zee's book is the easiest. Weyl, a great mathematician, wrote his book quite early in the game.

No references are given here for memory as a principle of nature. The references cited earlier should serve to capture many of the details.

MIND WITHOUT BIOLOGY: A CRITICAL POSTSCRIPT

Here I must give a rather long (but still incomplete) list arranged according to the order of the subheadings of the Critical Postscript.

PHYSICS: THE SURROGATE SPOOK

Penrose, R. *The Emperor's New Mind.* Oxford: Oxford University Press, 1989.

A very successful book in the sense of the size of its sales to the laity. Charming and lucid accounts of strange physics, quantum measurement, and so forth. But most of the book is practically irrelevant to its goals and claims about the mind, as I discuss in the text.

LOCKWOOD, M. *Mind, Brain, and the Quantum: The Compound "I."* Cambridge: Basil Blackwell, 1989.
A philosopher's discussion of many of the same issues covered by Penrose. Inconclusive.

ZOHAR, D. *The Quantum Self: Human Nature and Consciousness Defined by the New Physics.* New York: William Morrow, 1990.
Quantum this, quantum that, quantum everything. A book that goes as far out in the domain of physics as the surrogate spook as anything well-intentioned can go. Compared to Penrose, a very soft example of the genre.

DIGITAL COMPUTERS: THE FALSE ANALOGUE

HODGES, A. *Alan Turing: The Enigma.* New York: Simon & Schuster, 1983.
Biography of a remarkable mind and a life with a sad ending. Turing is the key theoretical figure in computing, outside of von Neumann. Of course, many other logicians and mathematicians set the stage, as I mention in the text.

JOHNSON-LAIRD, P. N. *The Computer and the Mind.* Cambridge: Harvard University Press, 1988.
The best account of the mind-as-machine view.

GRAUBARD, S. R., ed. "Artificial Intelligence." *Daedalus* 117, no. 1 (1988).
A series of essays on the subject, both supportive and critical.

PUTNAM, H. *Representation and Reality.* Cambridge: MIT Press, 1988.
A refutation of his own doctrine—Turing machine functionalism—by one of the most distinguished living philosophers.

ANDERSON, J. A., and E. ROSENFELD, eds. *Neurocomputing: Foundations of Research.* Cambridge: MIT Press, 1988. ANDERSON, J. A., A. PELLIONISZ, and E. ROSENFELD, eds. *Neurocomputing: Direction for Research.* Cambridge: MIT Press, 1990.
Two collections of papers on aspects of neural modeling and connectionism.

VICIOUS CIRCLES IN THE
COGNITIVE LANDSCAPE

GARDNER, H. *The Mind's New Science.* New York: Basic Books, 1985.
An excellent general survey of cognitive science.

WITTGENSTEIN, L. *Philosophical Investigations.* The English text of the 3rd ed. New York: Macmillan, 1953.
Already cited—pioneering in its early dissections of the problems of categorization and family resemblance. The article by G. C. Rota listed in the readings for chapter 14 pertains here too.

The second part of the figure on categorization and polymorphous sets (P–6, right) is from Dennis et al., "New problem in concept formation." *Nature* 243 (1973): 101–2.

ROSCH, E. "Human Categorization." In *Studies in Cross-Cultural Psychology*, ed. N. Warren. New York: Academic, 1977, 1–49.

An account by one of the most important psychologists in the area of categorization.

BERLIN, B., and P. KAY. *Basic Color Terms: Their Universality and Evolution*. Berkeley: University of California Press, 1969. TVERSKY, A., and D. KAHNEMAN. "Probability, Representativeness, and the Conjunction Fallacy." *Psychological Review* 90, no. 4 (1990): 293–315.

Pioneering studies on color categorization and on inference, inductive and otherwise.

The references to L. Rips and to L. Barsalou in the text may be found in their articles, which are included in *Similarity and Analogical Reasoning*, eds. S. Vosniadou and A. Ortony. Cambridge: Cambridge University Press, 1989.

FODOR, J. A. *Representations: Philosophical Essays on the Foundations of Cognitive Science*. Cambridge: MIT Press, 1981.

By the prolific philosopher and defender of "mentalese."

MARR, D. *Vision: A Computational Investigation into the Human Representation and Processing of Visual Information*. San Francisco: Freeman, 1982.

The last work by this late influential figure in psychophysics and neuroscience. Espouses the computational view but gives a good summary of "early" visual processes. Gets worse in later chapters as it addresses problems of categorization.

MILLIKAN, R. G. *Language, Thought, and Other Biological Categories: New Foundations for Realism*. Cambridge: MIT Press, 1984. "Thoughts Without Laws; Cognitive Science with Content." *Philosophical Review* XCV, no. 1 (Jan. 1986): 47–80.

A major figure in what I have called the Realists' Club, Millikan has put forth a powerful and original critique of what she calls meaning rationalism (roughly equivalent to what I have inveighed against in the Postscript).

GAULD, A. "Cognitive Psychology, Entrapment, and the Philosophy of Mind." In *The Case for Dualism*, eds. J. R. Smythies and J. Beloff. Charlottesville: University Press of Virginia, 1989, 187–253.

One does not have to agree with dualism to appreciate Gauld's scalding attack.

SHANON, B. "Semantic Representation of Meaning: A Critique." *Psychological Bulletin* 104, no. 1 (1988): 70–83.

A fine summary of the difficulties of functionalism, objectivism, and the ideas of mental representation.

PUTNAM, H. *Representation and Reality*. Cambridge: MIT Press, 1988.

Already mentioned, a chip from the master's workbench.

BRUNER, J. *Acts of Meaning*. Cambridge: Harvard University Press, 1990.

A beautiful essay by one of the founders of modern cognitive science pleading for the recognition of narrative as an important aspect of our mental life.

VON HOFSTEN, C. "Catching." In *Perspectives on Perception and Action*, ed. H. Heuer and A. F. Sanders. Hillsdale, N.J.: Lawrence Erlbaum Associates, 1987, 33–46.

An attack on the information processing view from the motor side.

LANGACKER, R. W. *Foundations of Cognitive Grammar*, vol. 1, *Theoretical Prerequisites.* Stanford: Stanford University Press, 1987.

An account by one of the early workers in this important field, work that has been extended by Lakoff (see the next reference).

LAKOFF, G. *Women, Fire, and Dangerous Things: What Categories Reveal About the Mind.* Chicago: University of Chicago Press, 1987. JOHNSON, M. *The Body in the Mind: The Bodily Basis of Meaning, Imagination, and Reason.* Chicago: University of Chicago Press, 1987.

Two important books by authors who have collaborated. They contain references to the other authors cited in this section. Johnson makes the case for metaphor as a powerful result of embodiment. Lakoff's more extensive account includes a history of ideas, a critique of the notion of mind without biology, and a plea for the recognition of embodiment as the basis for meaning and mind. It provides a basis for much of the last part of the Critical Postscript. It also describes at length the cognitive grammar I have summarized here. The present book and my trilogy may be considered a theoretical answer to the question *"How* is the mind embodied?" This question is raised by the aforementioned works in such a way that it cannot be ignored.

LANGUAGE: WHY THE FORMAL APPROACH FAILS

CHOMSKY, N. *Cartesian Linguistics.* New York: Harper & Row, 1966. *Rules and Representations.* New York: Columbia University Press, 1980.

Two works by the most influential linguist of recent times, a defender of the formal approach and its most powerful proponent.

LIGHTFOOT, D. *The Language Lottery: Toward a Biology of Grammars.* Cambridge: MIT Press, 1982.

An informative account by one of the epigones.

JACKENDOFF, R. *Consciousness and the Computational Mind.* Cambridge: MIT Press, 1987.

As good a summary of the combined view of language as syntax and mind as machine as one can get. The result: consciousness as an epiphenomenon. Obviously, I reject this view.

BRESNAN, J. ed. *The Mental Representation of Grammatical Relations.* Cambridge: MIT Press, 1982.

An extensive account of lexical functional grammar; one basis for the analysis by Pinker of language acquisition. Heavy going for neophytes.

PINKER, S. *Language Learnability and Language Development.* Cambridge: Harvard University Press, 1984.

An intelligent and penetrating analysis. Technical.

DONALDSON, M. *Children's Minds.* New York: Norton, 1978.

Less technical. Charming and yet nearly lethal in its attack on the idea of a language acquisition device.

LEVELT, W. J. M. *Speaking: From Intention to Articulation.* Cambridge: MIT Press, 1989. GRICE, H. P. *Logic and Conversation.* In *Studies in Syntax*, vol. 3, ed. P. Cole and J. L. Morgan. New York: Academic, 1967, 41–58.

Levelt's account deals with the actual problem of speaking in a comprehensive way. The framework is still mainstream cognitive science, however. Probably the best single volume on the subject. Grice is the highly original analyzer of the requirements for effective exchange in speech.

Vygotsky, L. S. *Thought and Language.* Cambridge: MIT Press, 1962.

The Soviet thinker (a colleague of A. Luria) who emphasized interpersonal and social exchanges and the "interiorization" of speech for purposes of thought.

Percy, W. *The Message in the Bottle: How Queer Man Is, How Queer Language Is, and What One Has to Do with the Other.* New York: Farrar, Straus & Giroux, 1976.

Amateur, moving, and at the same time, deep. Proof that thought requires no advanced degree and that having an M.D. is not always ruinous. Percy was a fine minor novelist.

Klima, E., and U. Bellugi. *The Signs of Language.* Cambridge: Harvard University Press, 1978.

A discussion of Bellugi's pioneering studies showing that sign language has a syntax, dialects, and other characteristics of spoken language.

Bickerton, D. *Language and Species.* Chicago: University of Chicago Press, 1990.

A valiant attempt to account for the evolution of speech by means of an intermediate "pidgin" or a protolanguage. Provocative.

Lieberman, P. *Uniquely Human: The Evolution of Speech, Thought, and Selfless Behavior.* Cambridge: Harvard University Press, 1991.

A good semipopular account by an authority on the evolution of the speech apparatus.

Keller, H. *The Story of My Life.* 1902. Reprint. New York: Doubleday, 1954.

Poignant. A fitting last reference—the achievement of linguistic behavior against all odds.

Credits

QUOTATIONS

Frontispiece from Empedocles, in J. Barnes, *The Presocratic Philosophers*, vol. 2 (New York: Routledge, Chapman and Hall, 1982), 180.
  Feynman, R., *The Character of Physical Law* (Cambridge: MIT Press, 1965), 125.

Dedication from Ecclesiastes, 6:18, circa 250 B.C.

Chapter 1 from Descartes, R., *Meditations*. In *The Philosophical Works of Descartes*, vols. 1 and 2, ed. E. Haldane and G. Ross (Cambridge: Cambridge University Press, 1978).
  De Unamuno, M., *Tragic Sense of Life*, trans. C. J. Flitch (New York: Macmillan, 1921), 34.

Chapter 2 from Whitehead, A. N., *Science and the Modern World* (New York: Macmillan, 1925), 2–3.

Chapter 3 from Maxwell, J. C., in *The Anthropic Cosmological Principle*, J. D. Barrow and F. J. Tipler (Oxford: Oxford University Press, 1986), 545.
  Diderot, D., *Rameau's Nephew/D'Alembert's Dream* (London: Penguin, 1966 [1769]), 210–12.

Chapter 4 from Adams, H., *The Education of Henry Adams* (New York: Houghton Mifflin, 1961).

Chapter 5 from Darwin, C., in H. Zinsser, *As I Remember Him: The Biography of R. S.* (Magnolia, Mass.: Peter Smith, 1970 [1939]).

CREDITS

Darwin, C., in *Charles Darwin's Notebooks 1836–1844: Geology, Transmutation of Species, Metaphysical Enquiries*, ed. P. H. Barrett, P. J. Gautrey, S. Herbert, D. Kohn, and S. Smith (Ithaca: Cornell University Press, 1987), 539.

Chapter 6 from Spitzer, N., "The Chicken and the Egg, Together at Last" (a review of *Topobiology: An Introduction to Molecular Embryology*), *The New York Times Book Review*, 22 January 1989, p. 12.

Chapter 7 from Pascal, B., *Pensées*, trans. A. J. Krailsheimer (New York: Penguin, 1966).

Chapter 8. Anonymous.

Chapter 9 from Crick, F.H.C., "Neural Edelmanism," *Trends in Neuroscience* 12, no. 7 (July 1989): 247.

Chapter 10 from Wittgenstein, L., in *Wittgenstein: The Philosophical Investigations*, ed. G. Pitcher (London: Macmillan, 1968), 465.
    Voltaire, in *Philosophical Dictionary*, vol. 1, ed. Peter Gay (New York: Basic Books, 1962), 308.

Chapter 11 from James, W., *Psychology: Briefer Course* (Cambridge: Harvard University Press, 1984), 401.

Chapter 12 from Focillon, H., in K. Atchity, *A Writer's Time*, (New York: Norton, 1986), 180.

Chapter 13 from Freud, S., *The Standard Edition of the Complete Psychological Works of Sigmund Freud*, vol. 14, trans. and ed. James Strachey (London: Hogarth Press, 1976), 280.
    Valéry, P., in W. H. Auden, *A Certain World* (New York: Viking, 1970), 260.

Chapter 14 from Bridgman, P. W., in "Quo Vadis" in *Daedalus* (Winter, 1958): 93.

Chapter 15 from Holmes, O. W., *The Complete Works of Oliver Wendell Holmes*. (St. Clair Shores, Mich.: Scholarly Press, 1972).
    Einstein, A., in K. Atchity, *A Writer's Time* (New York: Norton, 1986), 180.

Chapter 16 from Planck, M., in J. D. Barrow and F. J. Tipler, *The Anthropic Cosmological Principle* (Oxford: Oxford University Press, 1986), 123.
    Allen, W. By permission of Mr. Allen.

Chapter 17 from Schopenhauer, A., *Counsels and Maxims* (St. Clair Shores, Mich.: Scholarly Press, 1981).

Credits

Chapter 18 from Freud, S., *The Origins of Psycho-Analysis*, ed. M. Bonaparte, A. Freud, and E. Kris (New York: Basic Books, 1954), 120.

Chapter 19 from de la Mettrie, J. O., *L'Homme Machine*. Reprinted. (Peru, Ill.: Open Court Publishing, 1961), 140–41.

Chapter 20 from Einstein, A., in R. H. March, *Physics for Poets* (Chicago: Contemporary Books, 1978), 135.
Valéry, P., in *The Practical Cogitator*, ed. C. P. Curtis, Jr., and F. Greenslet (Boston: Houghton Mifflin, 1962), 597.

Critical Postscript quote in text from Quine, W. V., *The Ways of Paradox* (New York: Random House, 1966), 66.

ILLUSTRATIONS

Figures 1–1, 1–2, 4–1, and 5–1 from the Mary Evans Picture Library, London.

Figures 2–1 and 20–1 Copyright © American Institute of Physics. Reprinted by permission.

Figure 2–2 from H. Barrow, C. Blakemore, and M. Weston-Smith, ed., *Images and Understanding* (New York: Cambridge University Press, 1990), 261. Copyright © 1990 by Cambridge University Press. Reprinted by permission.

Figure 3–1 from C. Blakemore and S. Greenfield, ed., *Mindwaves* (New York: Basil Blackwell, 1987), 4. Copyright © 1987 by Basil Blackwell. Reprinted by permission.

Figures 3–2 and 3–3 from Gerald M. Edelman, *Topobiology: An Introduction to Molecular Embryology*. Copyright © 1988 by Basic Books, Inc. Reprinted by permission of HarperCollins Publishers.

Figure 3–5 (top left) from K. G. Pearson and C. S. Goodman, "Correlation of variability in structure with variability in synaptic connection of an identified interneuron in locusts," *Journal of Comparative Neurology* 184 (1979): 141–65. Reprinted by permission of Corey S. Goodman.

Figure 3–5 (bottom) from Michael M. Merzenich et al., "Topographic reorganization of somatosensory cortical areas 3b and 1 in adult monkeys following restricted deafferentation" and "Progression of change following median nerve section in the cortical representation of the hand in areas 3b and 1 in adult owl and squirrel monkeys," *Neuroscience* 8 (1983): 33–55 and *Neuroscience* (1983): 10:639–65. Reprinted by permission of Michael M. Merzenich.

Figure 3–5 (top right) from Eduardo R. Macagno, V. Lopresti, and C. Levinthal, "Structure and development of neuronal connections in isogenic organisms: variations and similarities in the optic system of *Daphnia magna*," *Proceedings of the National Academy of Science* 70 (1973): 57–61. Reprinted by permission of Eduardo R. Macagno.

Figure 4–1 (left) from W. and A. Durant, *The Story of Civilization: Rousseau and the Revolution*, vol. x (New York: Simon and Schuster, 1967). Reprinted by permission of the estate of Ethel Durant.

Figure 4–2 Reprinted by permission of Greenwood Publishing Group, Inc., Westport, Conn., from *Organization in Vision: Essays on Geshtalt Perception* by Gaetano Kanizsa (New York: Praeger, 1979), 78, and 74. Copyright © 1979 by Gaetano Kanizsa.

Figure 5–3 adapted from Richard Lewontin, *The Genetic Basis of Evolutionary Change* (New York: Columbia University Press, 1974). Copyright © 1974 by Richard Lewontin. Reprinted by permission.

Figure 5–4 (left) from Hugo Iltis, *The Life of Mendel* (New York: Hafner Publishing Co., 1966). Reprinted by Unwin Hyman, of HarperCollins Publishers Ltd.

Figure 5–5 reprinted from *Charles Darwin's Notebooks 1836–1844: Geology, Transmutation of Species, Metaphysical Enquiries*, transcribed and edited by Paul H. Barrett, Peter J. Gautrey, Sandra Herbert, David Kohn, and Sydney Smith. Originally published by the British Museum (Natural History). Copyright © 1987 by Paul H. Barrett, Peter Gautry, Sandra Herbert, David Kohn, Sydney Smith. Used by permission of the publisher, Cornell University Press.

Figure 5–6 from G. Ledyard Stebbins, *Darwin to DNA, Molecules to Humanity* (New York: W. H. Freeman and Company, 1982). Copyright © 1982 by W. H. Freeman and Company. Reprinted by permission.

Figure 6–2 from C. M. Anderson, F. H. Zucker, and T. A. Steitz, "Space-Filling Models of Kinase Clefts and Conformation Changes," *Science* 204 (1979): 375–80. Figure appears on p. 376. Copyright © 1979 by the American Association for the Advancement of Science. Reprinted by permission.

Figure 6–3 (top) from Gerald M. Edelman, "Cell Adhesion Molecules: A Molecular Basis for Animal Form," *Scientific American* 250, no. 4 (1984): 118–29. Figure appears on pp. 120–21. Copyright © 1984 by Scientific American, Inc. All rights reserved. Reprinted by permission.

Figure 6–3 (bottom) from Alfred Romer, *The Vertebrate Body*, 5th ed. (Orlando, Fla.: Saunders College Publishing, 1963), 119. Copyright © 1977 by Saunders College Publishing. Reprinted by permission of the publishers.

Figure P–3 (bottom) courtesy of N-Cube, Belmont, Calif.

Figure P–6 (right) from I. Dennis, J. A. Hampton, and S.E.G. Lea, "New Problem in Concept Formation," *Nature* 243 (1973): 101. Copyright © 1973 Macmillan Magazines, Ltd. Reprinted by permission of Ian Dennis and *Nature*.

Index

273